CRISPR Wars

How Gene Editing Will Revolutionize Warfare

By Sean Rust

©2024. All Rights Reserved.

Table of Contents

Introduction: The Dawn of a New Era

1. **CRISPR Unveiled**
2. **Military Aspirations and Genetic Engineering**
3. **Enhancing the Human Soldier**
4. **Ethical and Moral Dilemmas**
5. **The Geopolitical Landscape**
6. **The Science of Super-Soldiers**
7. **Tactical Advantages and Battlefield Realities**
8. **Public Perception and Media Influence**
9. **Cultural and Societal Impacts**
10. **Future Prospects and Speculations**

Conclusion: Preparing for a New Frontier

Introduction.
The Dawn of a New Era

Dr. Emily Carter, a renowned geneticist and professor at a prestigious American university, had dedicated her life to the peaceful advancement of science. Her groundbreaking research in CRISPR technology had earned her international acclaim. Little did she know, this very expertise would draw her into a covert military operation that threatened to unravel everything she believed in.

One cold winter evening, as Dr. Carter was wrapping up her work in the lab, a group of stern-faced men in military uniforms entered. Leading them was General Nathaniel Hayes, a high-ranking officer known for his ruthlessness and strategic brilliance. Without preamble, he presented her with an ultimatum: assist the military in a top-secret genetic modification program targeting U.S. Marines, or face dire consequences.

"Dr. Carter, your country needs you," General Hayes said, his voice devoid of any warmth. "We're preparing for a mission of utmost importance. You are to enhance our marines using your CRISPR technology. This is non-negotiable."

Dr. Carter's heart raced. The ethical implications of such a project were staggering. "General Hayes, my research is intended to cure diseases, not to turn soldiers into weapons. I cannot—"

"Spare me the lectures, Professor," Hayes interrupted. "This is about national security. We have intelligence suggesting that President Putin is planning a catastrophic attack against the United States. We must strike first, and we need every advantage we can get. Your work will ensure the mission's success."

The gravity of the situation was clear. Dr. Carter was left with no choice. Under heavy surveillance and constant pressure, she was relocated to a secluded military base where she began the harrowing task of genetically enhancing a select group of U.S. Marines. Using CRISPR, she edited their genes to increase muscle strength, accelerate healing, and enhance cognitive functions. Each day, she battled her conscience, knowing that her work could lead to unprecedented violence.

As the modifications took effect, the marines became formidable warriors, their abilities far surpassing any normal human's. General Hayes watched with satisfaction, his plans for the surprise attack on Moscow taking shape. The target was none other than President Putin, the leader of Russia and the perceived threat against the United States.

The night before the mission, Dr. Carter found herself alone in the lab, reflecting on the monstrous power she had helped create. She knew there was no turning back, but she also knew that such power could not be left unchecked. In a moment of quiet rebellion, she inserted a failsafe into the genetic modifications—a hidden sequence that could be activated to reverse the enhancements.

The day of the mission arrived. The genetically enhanced marines, now an elite force, embarked on their covert operation to infiltrate Moscow and neutralize Putin. As they moved with inhuman precision and strength, the plan seemed foolproof.

But Dr. Carter had sent an anonymous tip to an international watchdog organization, revealing the covert operation. The organization alerted global media, and within hours, the world was watching as the mission unraveled. International pressure and diplomatic interventions forced the U.S. government to abort the mission, averting what could have been a catastrophic escalation of hostilities.

In the aftermath, Dr. Carter was hailed as a whistleblower and a hero by some, and a traitor by others. The government disavowed the mission, and General Hayes was quietly retired. The incident sparked a global debate about the ethics of genetic modification in warfare, leading to stricter regulations and oversight.

Dr. Carter returned to her academic life, forever changed by the experience. She continued her research, but now with a renewed *focus on ensuring that her work would never again be used for harm. Her story served as a cautionary tale about the dangers of unchecked power and the ethical responsibilities of scientists in an age of unprecedented technological advancement.*

In a world where technological advancements are accelerating at an unprecedented pace, the potential for revolutionary changes in various fields is both exciting and daunting. One of the most profound developments in recent years is the advent of CRISPR technology, a tool that has the potential to transform not only medicine and agriculture but also the very nature of warfare.

CRISPR, which stands for Clustered Regularly Interspaced Short Palindromic Repeats, is a groundbreaking genetic editing technology discovered in the early 1990s. This tool allows scientists to make precise modifications to DNA, effectively rewriting the genetic code of living organisms. The implications of such a capability are vast and varied, offering the promise of eradicating genetic diseases, creating more resilient crops, and even altering the physical and cognitive traits of humans.

The potential to cure genetic diseases, improve agricultural yields, and advance scientific understanding has positioned CRISPR at the forefront of modern biotechnology. However, as with any powerful tool, the applications of CRISPR extend beyond the realm of medicine and agriculture, reaching into areas that spark profound ethical and societal debates.

Among the most controversial and impactful of these applications is the use of CRISPR technology in the military. The prospect of genetically enhanced soldiers has long been a topic of science fiction, but recent advancements have brought this idea closer to reality. Researchers and defense agencies are increasingly exploring how genetic modifications could provide tactical advantages in warfare. These developments raise critical questions about the ethical implications and the potential for misuse, echoing historical concerns about eugenics and the pursuit of human perfection.

As we consider the potential of CRISPR to revolutionize various aspects of human life, it is crucial to examine its implications for global security and military strategy. The intersection of genetic engineering and military technology presents a complex landscape, where the promise of enhanced capabilities must be weighed against the risks of unintended consequences and ethical dilemmas. Understanding the full scope of CRISPR's impact requires a thorough exploration of its potential applications in the context of defense and warfare.

The military applications of CRISPR are particularly significant. In the quest for enhanced soldiers, CRISPR offers the possibility of creating individuals with superior physical strength, heightened endurance, and accelerated healing abilities. Imagine a future where soldiers can operate in extreme environments without succumbing to fatigue or injury, where their cognitive functions are enhanced to make split-second decisions with unparalleled accuracy. These enhancements could give militaries a strategic advantage on the battlefield, reshaping the dynamics of global power.

The potential impact of CRISPR on warfare extends beyond individual enhancements. The technology could be used to develop biological weapons that target specific genetic traits, leading to a new era of biowarfare. These weapons could be designed to incapacitate or eliminate enemy forces while leaving allies unharmed, a prospect that raises significant ethical and security concerns.

As we stand on the brink of this new frontier, it is crucial to consider the broader implications of CRISPR technology. While the potential benefits are immense, the risks and ethical dilemmas are equally profound. The prospect of genetically modified soldiers and the weaponization of genetic technology present challenges that require careful consideration and robust regulatory frameworks.

The rapid advancement of CRISPR technology has sparked a revolution in genetic editing, bringing both hope and uncertainty. As this technology evolves, it holds the potential to cure genetic diseases and enhance human capabilities in ways previously unimaginable. However, the same technology that promises so much also raises profound ethical and societal questions. In a world where genetic editing becomes commonplace, what does it mean to be human? How far should we go in altering our genetic makeup, and who gets to decide what changes are acceptable?

The purpose of this book is to explore these critical questions by delving into the intersection of CRISPR technology and its potential military applications. By examining the scientific foundations, ethical dilemmas, and geopolitical implications, we aim to provide a comprehensive understanding of how genetic editing could reshape the future of warfare and society at large.

Our journey begins by unraveling the complexities of CRISPR technology and its capabilities. We will then investigate how militaries around the world are exploring genetic enhancements to create super-soldiers, capable of extraordinary feats. These developments are not without controversy, as they challenge our traditional notions of ethics and human rights.

Furthermore, this book seeks to shed light on the broader societal impact of genetic editing. As we navigate through the potential benefits and risks, it becomes clear that the quest for genetic perfection is fraught with peril. The history of eugenics serves as a stark reminder of the dangers associated with attempts to create a "perfect" population. We must consider the lessons of the past as we forge ahead into this uncharted territory.

In addition to the ethical considerations, the geopolitical landscape is poised for significant shifts. Nations may enter a new arms race, not with nuclear weapons, but with genetically enhanced soldiers. The implications for global stability and security are profound, necessitating a thorough examination of current international laws and the need for new regulatory frameworks.

Ultimately, this book aims to foster an informed and thoughtful dialogue about the future of genetic editing in the military context. By presenting a balanced view of the possibilities and challenges, we hope to engage readers in considering the complex interplay of science, ethics, and policy. The goal is not to provide definitive answers, but to encourage critical thinking and discussion about one of the most significant technological advancements of our time.

As we embark on this exploration, it is essential to keep an inquisitive yet serious tone. The potential for CRISPR technology to transform our world is immense, but so are the risks. By understanding both the science and the broader implications, we can better prepare for the future and make informed decisions about the role of genetic editing in our society.

Chapter 1.

CRISPR Unveiled

In the not-so-distant past, the idea of editing the very fabric of our genetic code seemed like a notion reserved for the realm of science fiction. Today, however, this idea has become a groundbreaking reality, largely thanks to the discovery of CRISPR technology. The journey of CRISPR from a curious observation in bacterial genomes to a revolutionary tool for genetic engineering is a fascinating tale of scientific ingenuity and perseverance.

The story of CRISPR begins in the late 1980s and early 1990s, when a relatively obscure geneticist named Francisco Mojica, working at the University of Alicante in Spain, stumbled upon an unusual repetitive DNA sequence in the genomes of archaea. These sequences, later termed "Clustered Regularly Interspaced Short Palindromic Repeats" (CRISPR), were initially a mystery. Mojica observed that these repetitive sequences were interspaced with unique segments of DNA, but their function remained elusive.

It wasn't until the mid-2000s that a clearer picture began to emerge. Researchers discovered that these unique DNA sequences were actually snippets of viral DNA, integrated into the bacterial genome as a form of immune defense. When a bacterium was attacked by a virus, it could transcribe these sequences into RNA, which would guide the Cas (CRISPR-associated) proteins to cut the viral DNA, thus neutralizing the threat. This adaptive immune system allowed bacteria to "remember" past infections and defend against future attacks more effectively.

The potential for this bacterial defense mechanism to be repurposed as a genetic editing tool was first recognized by a team led by Jennifer Doudna at the University of California, Berkeley, and Emmanuelle Charpentier at Umeå University in Sweden. In 2012, they published a seminal paper describing how the CRISPR-Cas9 system could be engineered to cut DNA at precise locations, guided by a custom RNA sequence. This breakthrough demonstrated that CRISPR-Cas9 could be used not just in bacteria, but in any organism, effectively transforming genetic engineering.

This discovery set off a cascade of research and development across the globe. Scientists quickly realized that CRISPR-Cas9 could be used to edit genes with unprecedented precision and efficiency, offering potential cures for genetic diseases, improvements in agricultural crops, and even applications in synthetic biology. The implications were vast, and the scientific community raced to explore the possibilities.

One of the most notable early applications of CRISPR technology was in the field of medicine. Researchers began using CRISPR to correct genetic mutations that cause diseases such as cystic fibrosis, muscular dystrophy, and sickle cell anemia.

In 2015, scientists successfully used CRISPR to repair a genetic defect in human embryos that causes a potentially fatal blood disorder, marking a significant milestone in the journey towards human gene therapy. This breakthrough occurred at Sun Yat-sen University in China, where a team led by Junjiu Huang targeted the HBB gene responsible for beta-thalassemia, a serious blood disorder. By precisely cutting out the defective part of the gene and replacing it with a healthy sequence, the researchers demonstrated the potential of CRISPR to correct inherited genetic disorders at the embryonic stage.

This experiment was pioneering for several reasons. Firstly, it showed that CRISPR could be used to edit the human germline, meaning the changes would be heritable and passed down to future generations. This opened up the possibility of eradicating genetic diseases from entire family lines. Secondly, the success of this procedure underscored the precision and effectiveness of CRISPR compared to older gene-editing techniques, which were less accurate and more prone to off-target effects.

However, this achievement also sparked intense ethical debates. The ability to edit human embryos raised concerns about the potential for "designer babies," where genetic modifications could be made for non-medical reasons, such as enhancing physical appearance or intelligence. The ethical implications of germline editing led to calls for stringent regulations and guidelines to ensure that this powerful technology is used responsibly. Despite these controversies, the 2015 experiment remains a landmark in the field of genetic engineering, illustrating both the profound potential and the significant challenges of CRISPR in human gene therapy.

Despite these ethical concerns, the momentum behind CRISPR research continued to grow. The technology was soon applied in agriculture, where it was used to create crops with enhanced nutritional profiles, resistance to pests and diseases, and improved yield. For example, scientists successfully used CRISPR to develop rice varieties that are resistant to bacterial blight, a devastating disease that can significantly reduce crop yields. By precisely targeting and editing the genes that make rice plants susceptible to the disease, researchers created strains that could thrive without the need for harmful chemical treatments.

Another significant application in agriculture involved using CRISPR to improve the nutritional content of crops. In 2016, researchers used CRISPR to enhance the levels of provitamin A in bananas, aiming to address vitamin A deficiency, which is a major health issue in many developing countries. This biofortification effort not only holds the promise of improving public health but also demonstrates the potential of CRISPR to make staple foods more nutritious.

CRISPR also found applications in environmental science, where it was used to develop genetically modified organisms capable of breaking down pollutants and combating the spread of invasive species. For instance, scientists have engineered bacteria to degrade plastic waste more efficiently. These bacteria can break down polyethylene terephthalate (PET), a common plastic used in bottles and packaging, into its constituent monomers, which can then be recycled into new plastic products. This innovation represents a significant step towards addressing the global plastic pollution crisis.

In another groundbreaking project, CRISPR was employed to tackle the problem of invasive species, such as the mosquito species Aedes Aegypti, which is responsible for spreading diseases like dengue fever, Zika virus, and chikungunya. By using CRISPR to modify the genes of these mosquitoes, researchers have developed strategies to reduce their populations. One approach involves creating genetically modified mosquitoes that produce non-viable offspring, effectively reducing the number of mosquitoes capable of transmitting these diseases.

These examples illustrate the wide-ranging potential of CRISPR technology beyond human health, highlighting its transformative impact on agriculture and environmental science. As CRISPR continues to evolve, its applications are likely to expand even further, offering innovative solutions to some of the world's most pressing challenges while also prompting ongoing discussions about the ethical implications of such powerful technology.

To truly appreciate the potential and complexities of CRISPR, it is important to understand its origins and natural function. The story of CRISPR begins with bacteria. In nature, bacteria use CRISPR as a defense mechanism against viruses. When a virus attacks a bacterium, the bacterium captures snippets of the virus's DNA and inserts them into its own genome in a specific pattern known as CRISPR. These sequences then act as a genetic memory bank, enabling the bacterium to recognize and defend against the virus in future encounters.

Central to the CRISPR system is the Cas9 protein, often described as molecular scissors. Cas9 can cut DNA at a precise location, guided by a piece of RNA called guide RNA (gRNA). This gRNA is designed to match the specific DNA sequence that needs to be edited. When Cas9 and gRNA are introduced into a cell, they seek out the matching DNA sequence in the cell's genome. Upon finding the target, Cas9 makes a cut in the DNA, effectively breaking the double-stranded helix.

What happens next depends on the cell's natural repair mechanisms. The cell can either try to repair the break by rejoining the cut ends, often introducing small errors that disrupt the gene, or by using a provided template to repair the break accurately, allowing for precise genetic modifications. This ability to cut DNA at specific sites and then either disrupt genes or insert new sequences is what makes CRISPR such a versatile and powerful tool.

To grasp the full potential of CRISPR, it's essential to understand the elegance of its simplicity. Unlike older gene-editing methods, which were time-consuming and imprecise, CRISPR offers a level of accuracy and efficiency previously unattainable. This technology allows for the editing of genes with unprecedented precision, making it possible to correct genetic defects, study gene functions, and even improve agricultural crops.

The process of editing genes using CRISPR can be broken down into several key steps. Designing the Guide RNA is the crucial first step in the CRISPR gene-editing process. This step involves creating a short RNA sequence, known as the guide RNA (gRNA), which is specifically tailored to match the DNA sequence that needs to be edited. The guide RNA is composed of two main parts: a scaffold sequence that binds to the Cas9 enzyme, and a spacer sequence that is complementary to the target DNA region.

To ensure precision, the spacer sequence of the gRNA must be unique and highly specific to the target gene. This specificity is vital because it ensures that the CRISPR-Cas9 system will only bind to and cut the intended DNA sequence, minimizing the risk of off-target effects that could potentially disrupt other genes and cause unintended consequences. Bioinformatics tools and databases are often used to design and validate these guide RNAs, allowing researchers to select the most effective and accurate sequences for their experiments.

Once the guide RNA is designed and synthesized, it can be paired with the Cas9 enzyme. This complex then acts as a molecular GPS, guiding the Cas9 to the exact location in the genome where the DNA cut is to be made. The precision of the guide RNA is what makes CRISPR-Cas9 such a powerful and versatile tool for gene editing, enabling scientists to modify genes with unprecedented accuracy and efficiency.

The next step in the CRISPR gene-editing process is to introduce the Cas9 protein and the guide RNA (gRNA) into the target cell. This precise delivery is crucial for the success of the gene-editing operation.

Typically, a vector is used to transport these components into the cell. Vectors are vehicles that can carry foreign genetic material into another cell. One common type of vector used for CRISPR delivery is a virus, which has been engineered to be non-pathogenic and can efficiently deliver the Cas9 protein and gRNA into the cell's nucleus. Viruses are adept at entering cells and integrating their genetic material into the host genome, making them effective tools for CRISPR delivery.

Apart from viral vectors, there are other methods to introduce CRISPR components into cells. Lipid nanoparticles, for example, are another delivery method where the CRISPR components are encased in lipid-based carriers that can fuse with cell membranes and release their payload inside the cell. Electroporation is another technique, which involves applying an electric field to cells to increase the permeability of the cell membrane, allowing the Cas9 protein and gRNA to enter.

Once inside the cell, the vector navigates to the nucleus, where the cell's DNA is housed. The Cas9 protein, guided by the gRNA, then locates the specific DNA sequence that needs to be edited. The precision with which the Cas9-gRNA complex identifies the target sequence is one of the defining features of CRISPR technology, allowing for highly accurate and efficient gene editing.

Introducing these components into the cell effectively and safely is critical for the success of CRISPR gene editing, whether the goal is to correct genetic defects, study gene functions, or develop genetically modified organisms. Each delivery method has its advantages and challenges, and the choice of method depends on factors such as the type of cell, the efficiency required, and the specific application of the gene-editing process.

When CRISPR-Cas9 creates a break in the DNA, the cell's natural repair mechanisms are activated to fix the damage. If the goal is to disable the gene, the cell employs a repair process called non-homologous end joining (NHEJ). This process attempts to rejoin the cut ends of the DNA strand. However, NHEJ is error-prone and often introduces small insertions or deletions at the break site, which can disrupt the gene's function and effectively disable it.

Alternatively, if the objective is to insert a new sequence of DNA, a different repair process called homology-directed repair (HDR) is used. In this case, scientists provide the cell with a template that contains the desired DNA sequence flanked by regions homologous to the sequences on either side of the break. The cell uses this template as a guide to accurately repair the break, incorporating the new sequence into its genome in the process. This allows for precise genetic modifications, such as correcting a mutation or inserting a new gene.

These natural repair mechanisms are essential for the functionality of CRISPR-Cas9 as a gene-editing tool, enabling scientists to either knock out genes to study their function or introduce specific genetic changes to investigate their effects or develop therapeutic interventions.

The precision and flexibility of CRISPR have opened up a world of possibilities for genetic research and therapy. However, it is important to note that the technology is still in its early stages, and there are many challenges to overcome. Issues such as off-target effects, where CRISPR makes unintended cuts in the genome, and ethical concerns about genetic modifications, particularly in humans, are areas of active research and debate.

The discovery of CRISPR technology has revolutionized the field of genetic engineering, offering unprecedented precision and versatility. This breakthrough has paved the way for numerous applications across various fields, each demonstrating the transformative potential of this genetic tool. Here are more examples of the effect of CRISPR on our world today.

One of the most significant applications of CRISPR is in the realm of medicine. CRISPR's ability to target and edit specific genes has opened up new possibilities for treating genetic disorders. For example, researchers have successfully used CRISPR to correct mutations responsible for diseases such as cystic fibrosis and sickle cell anemia in laboratory settings. In cystic fibrosis, a disease caused by mutations in the CFTR gene, CRISPR has been employed to repair the defective gene in patient-derived cells, demonstrating the potential to restore normal function. Similarly, in sickle cell anemia, which results from a single mutation in the HBB gene, CRISPR has been used to correct the mutation in hematopoietic stem cells, paving the way for the production of healthy red blood cells.

In 2020, a groundbreaking clinical trial showed promising results in treating sickle cell disease and beta-thalassemia using CRISPR-edited stem cells. In this trial, led by researchers from the Boston-based company Vertex Pharmaceuticals and CRISPR Therapeutics, patients with these debilitating blood disorders received an infusion of their own stem cells that had been genetically edited using CRISPR to produce fetal hemoglobin, a form of hemoglobin unaffected by the mutations causing their diseases. This approach effectively alleviated symptoms and reduced the need for regular blood transfusions.

The trial involved harvesting hematopoietic stem cells from the patients' bone marrow, editing these cells in the lab to reactivate the production of fetal hemoglobin, and then reinfusing the edited cells back into the patients. This not only demonstrated the potential of CRISPR for treating genetic blood disorders but also marked a significant milestone in the application of gene editing in clinical settings. The success of this trial has spurred further research and development, with the hope of expanding CRISPR-based therapies to a broader range of genetic diseases.

Moreover, CRISPR is being explored as a potential tool for combating viral infections, a frontier that could revolutionize the treatment of chronic diseases. Scientists have demonstrated CRISPR's ability to target and disable viral DNA within human cells, offering a novel and potentially curative approach to diseases that have long eluded effective treatments. For instance, research has shown promise in using CRISPR to excise HIV DNA from infected cells, effectively halting the replication of the virus and providing a potential pathway to eradicating the infection from the body. Similarly, studies have indicated that CRISPR can target and disrupt the DNA of hepatitis B virus (HBV) in liver cells, potentially curing a disease that affects hundreds of millions of people worldwide.

The method involves designing CRISPR-Cas9 systems that can precisely locate and cut viral DNA sequences integrated into the host genome. For HIV, this means identifying and cleaving the proviral DNA embedded in human cells, thus preventing the virus from hijacking the cellular machinery to produce new viral particles. Early studies have successfully demonstrated this in laboratory settings, where CRISPR has been used to remove HIV DNA from the genomes of infected cells, significantly reducing viral load and replication.

Similarly, hepatitis B research has focused on using CRISPR to target covalently closed circular DNA (cccDNA), a stable form of viral DNA that resides in the nucleus of infected liver cells. This cccDNA serves as a template for producing new viral particles, and its persistence is a major challenge in curing HBV. CRISPR's ability to disrupt cccDNA has shown potential in preclinical models, paving the way for new therapeutic strategies that could eliminate the virus from the liver and potentially cure the infection.

These advancements highlight the broad therapeutic potential of CRISPR in addressing some of the most persistent medical challenges. However, the approach is still in experimental stages and faces significant hurdles before it can be widely adopted in clinical practice. Challenges include ensuring the specificity and efficiency of CRISPR systems to avoid off-target effects that could cause unintended genetic alterations. Moreover, effective delivery mechanisms need to be developed to transport CRISPR components to the precise cells and tissues affected by the virus.

Regulatory and ethical considerations also play a critical role in the development and deployment of CRISPR-based therapies. As the technology advances, it is essential to establish rigorous standards for its use, ensuring that treatments are safe, effective, and ethically sound. The potential for off-target effects and unintended consequences must be thoroughly evaluated through extensive clinical trials before CRISPR can become a mainstream option for treating viral infections.

Despite these challenges, the progress made thus far underscores the transformative potential of CRISPR. As research continues to advance, there is hope that CRISPR could one day provide cures for diseases that have plagued humanity for decades, if not centuries. The ongoing exploration of CRISPR for combating viral infections represents a beacon of innovation, promising new avenues for medical science and the potential to improve the lives of millions of people around the world.

In agriculture, CRISPR technology is being harnessed to improve crop resilience, yield, and nutritional value. By editing the genomes of plants, scientists are developing crops that can withstand harsh environmental conditions, resist pests, and reduce the need for chemical pesticides. For instance, researchers have used CRISPR to create rice varieties that are more resistant to bacterial blight, a significant threat to rice production worldwide. Bacterial blight can cause severe yield losses, and traditional methods of control often involve extensive use of chemicals. The CRISPR-edited rice plants have been engineered to carry a mutation in the SWEET gene family, which renders them less susceptible to the pathogen, thereby reducing the reliance on chemical treatments and increasing overall crop productivity.

In addition to disease resistance, CRISPR is being used to develop crops that can tolerate extreme weather conditions. This is increasingly crucial as climate change makes weather patterns more unpredictable and severe. For instance, scientists have successfully modified the genome of wheat to improve its tolerance to drought and heat. These modifications are essential because wheat is a staple food crop globally, and its yield stability is vital for food security.

To achieve this, researchers focus on specific genes involved in the plant's stress response pathways. One notable example is the modification of genes that regulate stomatal closure and water retention. By enhancing these traits, scientists have developed wheat varieties that can conserve water more efficiently during drought conditions, thus maintaining their growth and productivity.

Another approach involves altering the expression of genes related to heat shock proteins (HSPs), which help protect plant cells from the damaging effects of high temperatures. Enhanced expression of these proteins enables the wheat plants to better survive and continue to produce grain during periods of extreme heat. This genetic modification ensures that the plants' metabolic processes remain stable, reducing the risk of heat-induced damage and yield loss.

Moreover, researchers have targeted genes responsible for root growth and development. By promoting deeper and more robust root systems, modified wheat plants can access water and nutrients from deeper soil layers, providing a buffer against surface-level drought. This adaptation not only helps the plants survive in dry conditions but also improves soil health by preventing erosion and maintaining soil structure.

These advancements in crop genetic engineering are not limited to wheat. Similar techniques are being applied to other essential crops like rice, maize, and soybeans. For example, CRISPR has been used to develop rice varieties with improved salt tolerance, allowing them to thrive in saline soils where traditional rice strains would fail. This is particularly important for regions where rising sea levels and soil salinization threaten agricultural productivity.

The use of CRISPR to enhance crop resilience to extreme weather conditions demonstrates the technology's potential to address some of the most pressing challenges in agriculture today. By ensuring that crops can withstand the stresses imposed by climate change, scientists are working to safeguard food supplies and support the livelihoods of farmers worldwide. As research continues, the integration of CRISPR technology in agriculture promises to contribute significantly to global food security and sustainable farming practices.

CRISPR is also being used to enhance the nutritional content of crops. A notable example is the biofortification of tomatoes to increase their vitamin C content, which could have significant health benefits in regions where nutritional deficiencies are prevalent. Researchers have successfully edited the genes responsible for ascorbic acid (vitamin C) biosynthesis in tomatoes, resulting in varieties with significantly higher levels of this essential nutrient. This biofortification effort is particularly important for populations that rely heavily on staple crops and have limited access to a diverse diet.

Another example of nutritional enhancement through CRISPR is the development of maize varieties with increased levels of provitamin A. This effort aims to combat vitamin A deficiency, which is a major cause of preventable blindness and immune deficiencies in developing countries. By editing the carotenoid biosynthetic pathway, scientists have created maize with higher beta-carotene content, which the body can convert into vitamin A.

These advancements in agricultural biotechnology not only promise to enhance food security but also address global nutritional challenges. By reducing the need for chemical inputs, increasing resilience to climate change, and improving the nutritional value of staple crops, CRISPR technology offers a sustainable approach to meeting the growing food demands of the world's population. As research progresses, the potential for

CRISPR to revolutionize agriculture and contribute to global health and environmental sustainability becomes increasingly evident.

CRISPR's potential extends beyond human health and agriculture to environmental conservation. One of the most innovative applications in this field is the concept of gene drives, which use CRISPR to spread specific genetic traits through populations of species at an accelerated rate. This technology holds promise for controlling invasive species and disease vectors that threaten ecosystems and human health.

Gene drives work by biasing the inheritance of a particular gene so that it is passed on to nearly all offspring, rather than the typical 50% inheritance rate seen in traditional genetics. This ensures that the targeted genetic trait spreads rapidly through the population. Researchers are exploring several groundbreaking applications of gene drives to address some of the most pressing environmental and public health challenges.

For example, scientists are investigating the use of gene drives to reduce populations of malaria-carrying mosquitoes, specifically the Anopheles species. Malaria is a devastating disease that affects millions of people annually, primarily in sub-Saharan Africa. By introducing a gene drive that either renders female mosquitoes infertile or significantly reduces their lifespan, researchers aim to drastically reduce the mosquito population and, consequently, the incidence of malaria. Initial laboratory studies have shown promising results, with gene drives rapidly spreading through mosquito populations and achieving significant reductions in their numbers.

Another application of gene drives is in the control of invasive species that disrupt local ecosystems. For instance, the invasive cane toad in Australia has caused severe ecological damage since its introduction in the 1930s. Researchers are exploring gene drives that could either reduce the fertility of cane toads or increase their susceptibility to certain diseases, thereby controlling their population and mitigating their impact on native wildlife.

Additionally, gene drives are being studied for their potential to preserve endangered species. For example, they could be used to introduce genetic traits that confer resistance to diseases threatening particular species, such as the white-nose syndrome in bats or chytridiomycosis in amphibians. By enhancing the resilience of these species, gene drives could play a critical role in conservation efforts.

The potential benefits of gene drives are significant, but they also come with considerable ethical and ecological concerns. The irreversible nature of gene drives and their ability to spread rapidly through populations raise questions about unintended consequences and the long-term impact on ecosystems. Therefore, extensive research, careful planning, and stringent regulatory frameworks are essential to ensure the responsible use of this powerful technology in environmental conservation.

However, the use of gene drives also raises ecological and ethical concerns, as the long-term impacts on ecosystems are not yet fully understood. This highlights the importance of careful consideration and regulation in the application of CRISPR technology for environmental purposes.

In the field of industrial biotechnology, CRISPR is being utilized to optimize the production of biofuels and biochemicals. By modifying the genomes of microorganisms, scientists are enhancing their ability to produce biofuels more efficiently, offering a sustainable alternative to fossil fuels. For example, CRISPR has been used to engineer yeast strains that can ferment biomass into ethanol more effectively, thereby improving the efficiency of biofuel production. Specifically, researchers have targeted and modified genes in yeast to increase their tolerance to ethanol and enhance their ability to convert various types of biomass, including agricultural residues and woody materials, into fermentable sugars. This genetic optimization not only boosts ethanol yield but also reduces the costs and environmental impact of biofuel production processes.

Additionally, CRISPR is aiding in the development of microorganisms that can produce valuable biochemicals, such as pharmaceuticals and industrial enzymes. For instance, CRISPR has been employed to modify bacterial strains like Escherichia coli and Streptomyces, enabling them to synthesize complex antibiotics and anticancer agents more efficiently. This genetic manipulation allows for higher yields and purity of these critical drugs, reducing the dependency on traditional extraction methods from natural sources, which can be labor-intensive and environmentally damaging.

Moreover, in the production of industrial enzymes, CRISPR has facilitated the creation of enzyme variants with improved stability, activity, and specificity. These enzymes are used in a wide range of applications, from detergents and food processing to bioremediation and paper manufacturing. For example, by editing the genome of Aspergillus niger, a common industrial fungus, scientists have enhanced its ability to produce high levels of phytase, an enzyme crucial for animal feed that breaks down phytic acid, improving the nutritional availability of phosphorus in livestock diets.

These advancements demonstrate how CRISPR technology is revolutionizing industrial biotechnology, making the production of biofuels and biochemicals more efficient, cost-effective, and environmentally friendly. By leveraging the power of genetic editing, industries can transition towards more sustainable practices, reducing their carbon footprint and contributing to a greener economy.

CRISPR's influence has even reached the cosmetic industry, where it is being explored for its potential to revolutionize skin care and anti-aging treatments. By targeting genes associated with skin aging and pigmentation, CRISPR offers the possibility of developing more effective and long-lasting cosmetic treatments. For instance, researchers are investigating the use of CRISPR to correct genetic mutations that lead to skin disorders or accelerate the aging process.

One promising area of research involves using CRISPR to target and edit genes responsible for collagen production. Collagen is a protein that provides structure and elasticity to the skin, but its production naturally decreases with age, leading to wrinkles and sagging skin. By enhancing the expression of genes involved in collagen synthesis, CRISPR could help maintain higher levels of collagen, resulting in firmer, more youthful skin.

Another exciting application is in the treatment of hyperpigmentation disorders, such as melasma and age spots. These conditions are often caused by the overproduction of melanin, the pigment that gives skin its color. Scientists are exploring how CRISPR can be used to regulate the genes that control melanin production, potentially offering a way to reduce unwanted pigmentation and achieve a more even skin tone.

CRISPR is also being studied for its potential to address genetic conditions that affect the skin, such as epidermolysis bullosa (EB), a group of rare diseases that cause the skin to become very fragile and blister easily. By correcting the genetic mutations that cause EB, CRISPR could provide a permanent treatment for this debilitating condition, improving the quality of life for those affected.

Additionally, the anti-aging industry is looking into the use of CRISPR to delay or even reverse the cellular aging process. Researchers are identifying genes that play a role in cellular senescence, the process by which cells stop dividing and contribute to aging and age-related diseases. By editing these genes, it may be possible to rejuvenate cells and extend their healthy lifespan, offering new approaches to anti-aging therapies.

The potential of CRISPR in the cosmetic industry is vast, promising not only to enhance the effectiveness of skin care products but also to bring about groundbreaking treatments that go beyond surface-level improvements, addressing the genetic foundations of skin health and aging. As research continues, the integration of CRISPR into cosmetic science could lead to personalized, gene-based treatments that cater to individual genetic profiles, ushering in a new era of customized skin care solutions.

While these applications are still in their early stages, they represent the wide-reaching impact of CRISPR technology across various sectors. The ability to precisely edit genetic material holds promise for innovations that were once thought to be the realm of science fiction.

As we delve deeper into the military aspirations and genetic engineering in the following chapters, it's crucial to understand the foundational capabilities of CRISPR that make such advancements possible. This understanding sets the stage for exploring how these capabilities could be harnessed—or misused—in the context of modern warfare and beyond.

Chapter 2

CRISPR Unveiled

The quest to enhance human capabilities for military purposes is not a new phenomenon. Throughout history, nations have sought various means to give their soldiers an edge on the battlefield. From ancient herbal concoctions to modern pharmacological solutions, the drive to push human limits has been relentless. The historical context of these enhancements provides crucial insights into the ethical, strategic, and practical considerations that accompany the adoption of genetic technologies like CRISPR in contemporary military settings.

In ancient times, warriors consumed various natural substances believed to enhance their physical and mental abilities. For example, the ancient Greeks and Romans used opium and other herbal mixtures to dull pain and boost endurance. Opium, derived from the poppy plant, was known for its potent analgesic properties and was commonly used to relieve pain in various medical treatments. The Greek physician Hippocrates, often referred to as the "Father of Medicine," documented the use of opium in his medical texts, highlighting its effectiveness in treating ailments ranging from headaches to respiratory issues. In addition to opium, the Greeks and Romans utilized a variety of herbal concoctions to enhance physical performance and endurance. One notable example is the use of the herb silphium, which was highly prized in the ancient world for its medicinal properties. Silphium was believed to have numerous benefits, including pain relief, fever reduction, and even contraceptive effects. It was so valuable that it was depicted on ancient coins and traded extensively across the Mediterranean.

The berserkers of Norse legend reportedly ingested hallucinogenic mushrooms to induce a state of fierce, uncontrollable rage during battle. These warriors, known for their frenzied fighting style and seemingly superhuman strength, were feared and revered in equal measure. The term "berserker" itself is derived from the Old Norse words "berr" (bear) and "serkr" (shirt), suggesting that these warriors either wore bear skins into battle or fought without armor, embodying the ferocity of a wild animal.

Historical accounts and sagas describe berserkers as entering a trance-like state, often attributed to the consumption of Amanita muscaria, a type of hallucinogenic mushroom commonly found in the forests of Northern Europe. This mushroom contains psychoactive compounds like muscimol and ibotenic acid, which can cause intense hallucinations, altered perceptions of reality, and increased aggression. Under the influence of these substances, berserkers would reportedly become impervious to pain and capable of extraordinary feats of strength, attacking their enemies with relentless fury and little regard for their own safety.

This altered state, sometimes referred to as "berserkergang," was not just a result of the mushrooms but was also likely influenced by ritualistic practices, psychological conditioning, and the intense cultural significance placed on martial prowess and fearlessness in battle. The combination of these factors created a powerful psychological and physical transformation, enabling berserkers to perform acts that seemed beyond the capabilities of ordinary men.

While the exact historical accuracy of these accounts is debated, the concept of the berserker has left a lasting impact on both historical and modern perceptions of Norse warriors. The image of the unstoppable, mushroom-fueled warrior has permeated popular culture, appearing in literature, films, and even video games, where berserkers are often depicted as the ultimate embodiment of raw, untamed power.

Beyond the legends, the use of psychoactive substances to enhance combat performance is not unique to the Norse berserkers. Similar practices have been recorded in various cultures throughout history, from the use of coca leaves by Incan warriors to the consumption of hashish by the fabled Assassins of the Middle East. These historical parallels highlight a broader human fascination with the potential to transcend ordinary physical and mental limits through the use of natural substances, a theme that continues to capture the imagination of people today.

These early attempts at enhancement, though rudimentary, laid the groundwork for more systematic approaches in later periods. These natural substances, while somewhat effective, often lacked consistency and had unpredictable side effects. The trial and error nature of these methods highlighted the need for more controlled and reliable means of enhancement.

In the 1920s, the Soviet Union embarked on a bizarre and controversial scientific endeavor led by the prominent biologist Ilya Ivanov. Known for his work in animal breeding, Ivanov aimed to push the boundaries of science by creating a human-ape hybrid. This initiative, fueled by the Soviet regime's desire to demonstrate the superiority of socialism through scientific achievement, was both ambitious and ethically dubious. One of the underlying purposes of this research was to develop a new kind of soldier—one with the strength and resilience of an ape combined with the intelligence and obedience of a human. The Soviet leadership envisioned these hybrids as formidable assets in warfare, capable of performing tasks beyond the physical limits of ordinary soldiers.

In 1926, Ivanov traveled to French Guinea to conduct his experiments, where he attempted to inseminate female chimpanzees with human sperm. Despite his efforts, these initial experiments failed to produce any viable results. Undeterred, Ivanov returned to the Soviet Union and continued his work with the support of the Soviet government. This support reflected the broader Soviet ideology that saw science as a means to achieve extraordinary feats and surpass the capitalist West, including the potential militarization of biological advancements.

Back in the Soviet Union, Ivanov shifted his focus to using ape sperm to inseminate human women, a plan that raised significant ethical and moral concerns. He received permission from Soviet authorities to proceed, but finding willing female participants proved difficult. Eventually, Ivanov's experiments were shut down by the Soviet government, which had begun to view his work as an embarrassment and a distraction from other scientific pursuits. The impracticality and ethical issues surrounding the project ultimately led to its abandonment.

The scandal surrounding Ivanov's experiments highlighted the ethical boundaries that were being pushed in the name of scientific progress and military advantage. It also underscored the Soviet Union's intense desire to achieve scientific supremacy, even at the cost of conducting morally questionable research. Despite his failure, Ivanov's work left a lasting impact on the scientific community, serving as a cautionary tale about the potential dangers of unfettered scientific experimentation without ethical oversight.

This peculiar chapter in Soviet history remains a fascinating example of how ideological and military ambitions can drive scientific endeavors to extreme lengths, often ignoring the ethical implications. Ivanov's experiments, while ultimately unsuccessful, continue to spark discussions about the limits of scientific research, the intersection of science and warfare, and the moral responsibilities of scientists.

The industrial revolution brought significant advancements in medicine and chemistry, leading to the development of synthetic drugs that could be used to enhance soldier performance. During World War II, both the Allies and the Axis powers employed amphetamines to keep troops alert and combat fatigue. German soldiers were given Pervitin, a type of methamphetamine, which was believed to increase their stamina and aggressiveness. Similarly, the Allies used Benzedrine, another amphetamine, to help pilots and infantrymen stay awake during long missions. These drugs, while effective in the short term, often had severe side effects and led to addiction, highlighting the dangers of chemical enhancements. The transition to synthetic drugs represented a significant improvement in terms of potency and reliability compared to natural substances. However, the long-term health implications and potential for abuse underscored the need for safer alternatives.

The Cold War era saw further advancements in military enhancement programs, driven by the intense competition between the United States and the Soviet Union. Both superpowers invested heavily in research to improve the physical and cognitive capabilities of their soldiers. The U.S. military explored various performance-enhancing drugs and techniques, including the use of steroids to build muscle mass and strength. The Soviet Union, on the other hand, experimented with a range of substances, including phentermine and various psychotropic drugs, to boost endurance and mental resilience. These enhancements aimed not only at improving physical capabilities but also at maintaining mental sharpness and resilience under extreme stress. The Cold War era's innovations were marked by a more scientific and experimental approach, reflecting advancements in understanding human physiology and psychology.

One notable example from the Cold War period is the U.S. Army's "Edgewood Arsenal experiments," which took place from 1948 to 1975. These experiments involved the testing of various chemical agents on military personnel to evaluate their potential for enhancing performance and inducing incapacitation in enemies. The substances tested included hallucinogens, nerve agents, and stimulants. Among the hallucinogens tested was LSD, which was administered to soldiers to study its effects on their mental state and performance. The use of nerve agents, such as sarin and VX, was also explored to understand their potency and the potential for protective measures against chemical warfare.

The stimulants tested included compounds like amphetamines, which were examined for their ability to enhance alertness and endurance in soldiers. The results of these tests varied, with some substances proving effective in achieving their intended goals, while others caused severe and unpredictable side effects.

While some of the findings contributed to the development of new military technologies and defensive measures, the ethical implications of these experiments were profound. Many of the soldiers involved were not fully informed of the nature or potential risks of the tests, raising serious concerns about informed consent. The long-term health effects on participants were often overlooked or inadequately addressed, leading to significant suffering and health issues for many involved.

The ethical breaches and lack of transparency eventually led to public outcry and demands for accountability. In response to growing awareness and criticism, the U.S. government implemented stricter regulations on human experimentation. The aftermath of the Edgewood Arsenal experiments prompted the establishment of the Belmont Report in 1979, which laid out fundamental ethical principles and guidelines for conducting research involving human subjects.

The Edgewood Arsenal experiments demonstrated the willingness of military institutions to push ethical boundaries in the pursuit of enhanced capabilities, ultimately prompting a reevaluation of the moral limits of such research. The legacy of these experiments serves as a cautionary tale, highlighting the importance of ethical considerations and the protection of human rights in scientific and military research. The lessons learned from this period have informed contemporary debates on the ethical implications of emerging technologies, including CRISPR and genetic engineering, underscoring the need for vigilance and ethical oversight in the pursuit of scientific and technological advancements.

In more recent times, advancements in biotechnology have opened new avenues for military enhancements. The concept of "biosoldiers" emerged, focusing on using biological means to improve human capabilities. This includes the exploration of gene therapy, prosthetics, and other bioengineering techniques. For instance, DARPA (Defense Advanced Research Projects Agency) has been at the forefront of research into augmenting human performance through biological means.

Their projects have included developing exoskeletons to enhance physical strength and endurance, as well as investigating ways to improve cognitive function and resilience to stress. The development of exoskeletons, for instance, has been a significant focus for enhancing the physical capabilities of soldiers. These wearable machines are designed to augment human strength and endurance, allowing soldiers to carry heavy loads with less fatigue and reducing the risk of injury. Advanced exoskeletons can provide support to joints and muscles, enabling soldiers to move more efficiently over long distances and in challenging terrains. Companies like Lockheed Martin and DARPA have made significant strides in creating exoskeletons that are lightweight, flexible, and capable of being integrated seamlessly with the body's movements.

In addition to physical enhancements, there has been substantial research into improving cognitive function and resilience to stress. Cognitive enhancements aim to sharpen mental acuity, improve decision-making abilities, and increase overall brain function. Researchers are exploring various methods to achieve these goals, including neurostimulation techniques such as transcranial direct current stimulation (tDCS) and transcranial magnetic stimulation (TMS). These methods involve non-invasive procedures that use electrical currents or magnetic fields to stimulate specific areas of the brain, potentially enhancing cognitive performance and focus.

The integration of technology and biology also extends to the development of pharmacological solutions aimed at boosting cognitive function. Nootropics, or "smart drugs," are substances that can enhance memory, creativity, and motivation in healthy individuals. Military research has investigated the use of these drugs to improve soldiers' cognitive performance, particularly in high-stress and demanding environments. For example, modafinil, a drug initially developed to treat narcolepsy, has been used to enhance alertness and cognitive function in soldiers who need to stay awake and focused for extended periods.

Improving resilience to stress is another critical area of focus. Military personnel are often exposed to extreme stressors that can affect their mental health and performance. Research in this field includes the study of stress inoculation training, which involves exposing individuals to stress in controlled settings to build their resilience. Additionally, there is ongoing investigation into the genetic and molecular mechanisms that underlie stress responses. By understanding these mechanisms, scientists hope to develop interventions that can enhance stress resilience at a biological level.

Among DARPA's notable projects is "Project Avatar," an ambitious initiative that seeks to develop technologies enabling soldiers to remotely control semi-autonomous bipedal machines. This project represents a significant leap forward in integrating advanced robotics with neural interfaces, potentially revolutionizing the way military operations are conducted. By allowing soldiers to perform dangerous tasks without physical risk, Project Avatar aims to enhance both the safety and effectiveness of military personnel.

The core concept of Project Avatar revolves around the creation of robotic surrogates that can be controlled by human operators through a combination of wearable sensors and brain-computer interfaces (BCIs). These surrogates, or avatars, are designed to mimic human movement and capabilities, equipped with dexterous limbs, advanced sensory systems, and sophisticated AI for autonomous functions. The integration of neural interfaces allows soldiers to control these avatars intuitively, using their thoughts to direct the machines' actions in real-time.

One of the primary goals of Project Avatar is to reduce the physical risk to soldiers by enabling them to engage in hazardous missions remotely. This could include tasks such as bomb disposal, reconnaissance in hostile environments, and frontline combat operations. By deploying robotic avatars in place of human soldiers, military units can maintain a presence in dangerous areas without exposing personnel to life-threatening situations. This approach not only enhances soldier safety but also allows for greater operational flexibility and efficiency.

The development of neural interfaces for Project Avatar is a critical aspect of its success. These interfaces rely on cutting-edge research in neuroscience and bioengineering to translate neural signals from the human brain into commands for the robotic avatar. Techniques such as electroencephalography (EEG) and intracortical microelectrode arrays are being explored to achieve precise and reliable control over the avatars. The challenge lies in creating interfaces that are both highly responsive and minimally invasive, ensuring that soldiers can operate the avatars seamlessly without significant physical or cognitive strain.

Project Avatar also underscores the military's broader interest in integrating advanced biological and technological enhancements. While not strictly a genetic modification project, it aligns with DARPA's overarching goal of augmenting human capabilities through technology. This includes initiatives in areas such as augmented reality, exoskeletons, and cybernetic implants, all aimed at creating a more resilient and capable soldier.

The potential applications of Project Avatar extend beyond traditional military operations. In disaster response scenarios, avatars could be deployed to assess damage, rescue survivors, and perform critical repairs in environments that are too dangerous for human responders. In law enforcement, robotic avatars could be used for surveillance, hostage negotiations, and crowd control, reducing the risk to officers and civilians alike.

However, the development of such advanced technologies also raises important ethical and practical questions. The use of remote-controlled avatars in combat scenarios could blur the lines of accountability and decision-making, raising concerns about the potential for misuse and unintended consequences. Additionally, the psychological impact on soldiers who control these avatars, particularly in high-stress and combat situations, requires careful consideration and support.

Furthermore, the integration of neural interfaces and robotic systems poses significant technical challenges. Ensuring robust and secure communication between the human operator and the avatar is crucial to prevent hacking or interference. The avatars themselves must be designed to withstand the rigors of combat and operate reliably in diverse and unpredictable environments.

Despite these challenges, Project Avatar represents a bold step toward the future of military technology. By merging advanced robotics with neural interfaces, DARPA aims to create a new paradigm in which soldiers can extend their reach and capabilities far beyond their physical limitations. This project exemplifies the innovative spirit of DARPA and its commitment to pushing the boundaries of what is possible, ultimately striving to enhance the safety and effectiveness of military operations while exploring the profound implications of human-machine integration.

These advancements represent a shift towards integrating technology and biology in ways that previous methods could not achieve, offering more precise and potentially safer enhancements. The convergence of fields such as biotechnology, neuroscience, and robotics is enabling the development of solutions that can be tailored to the specific needs of individuals. For instance, personalized medicine approaches can be employed to create customized enhancement plans based on a soldier's genetic makeup, lifestyle, and specific operational requirements.

Furthermore, the ethical and safety concerns associated with these advancements are being carefully considered. Unlike earlier methods that often had significant side effects or ethical issues, modern enhancements are designed with a greater emphasis on safety and ethical standards. For example, the development of exoskeletons includes rigorous testing to ensure they do not cause harm or undue strain on the user's body.

Similarly, cognitive enhancement techniques are being evaluated for their long-term effects and potential risks to ensure they are safe for widespread use.

The potential applications of these enhancements extend beyond the military. In civilian life, exoskeletons could be used to assist individuals with mobility impairments, providing them with greater independence and improved quality of life. Cognitive enhancement techniques and stress resilience training could benefit professionals in high-stress occupations, such as first responders, healthcare workers, and executives, helping them to perform better under pressure and maintain their mental well-being.

The allure of genetic modification lies in its potential to create super-soldiers with innate advantages that surpass traditional methods of enhancement. Unlike external tools and temporary measures, genetic engineering can alter the very blueprint of human biology, embedding enhancements directly into a soldier's DNA. This shift from external to internal modification reflects a broader trend in military science: the pursuit of permanent, reliable, and heritable traits that can be passed down through generations. This transition also signifies a move towards a more holistic approach to soldier enhancement, where the focus is not just on physical capabilities but also on cognitive and psychological resilience.

CRISPR technology has revolutionized genetic research, offering unprecedented precision in editing DNA. For military applications, CRISPR presents an opportunity to create genetically modified soldiers who are stronger, faster, and more resilient. Research is ongoing into how CRISPR can be used to enhance muscle growth, improve cognitive functions, and increase resistance to environmental stresses such as extreme temperatures and radiation.

The potential of genetic technology in the military has not gone unnoticed by other nations. Countries like China and Russia are reportedly investing heavily in similar research, potentially sparking a new kind of arms race. The Chinese military, for example, has expressed significant interest in biotechnology, viewing genetic engineering as a critical component of future warfare.

This international competition raises several ethical and regulatory concerns. The prospect of creating "designer soldiers" brings up questions about consent, human rights, and the potential for abuse. Unlike traditional weapons, genetic modifications have permanent and far-reaching implications for individuals and society.

The path to deploying genetically modified soldiers is fraught with scientific, ethical, and practical challenges. Scientifically, the long-term effects of genetic modifications are still unknown. Ethical concerns include the potential for coercion, the impact on individual identity, and the broader societal implications of creating a class of enhanced humans.

Practically, there are significant hurdles in terms of regulation and oversight. International laws and treaties, such as the Biological Weapons Convention, do not currently cover genetic modifications, creating a regulatory grey area. Additionally, public perception and acceptance of genetically modified soldiers are uncertain, with potential backlash from various societal and cultural groups.

Despite these challenges, the military's interest in genetic technology is unlikely to wane. The potential advantages are too significant to ignore. However, it is crucial that this interest is balanced with careful consideration of the ethical, legal, and social implications. As we delve deeper into the science and potential applications of CRISPR in the military, we must keep in mind the broader context. The next section will explore specific case studies and expert opinions, providing a more detailed look at the real-world applications and challenges of these technologies.

As the discussion about the military's fascination with genetic engineering continues, it is impossible to ignore the plethora of conspiracy theories that have emerged over the years. These theories, while often sensational and unverified, reflect a deep-seated public concern about the potential misuse of genetic technologies in creating super-soldiers.

One of the most persistent conspiracy theories revolves around secret government programs allegedly dedicated to developing super-soldiers. These theories suggest that various military organizations worldwide have been experimenting with genetic modifications for decades.

In recent years, the rise of CRISPR technology has fueled a new wave of conspiracy theories. CRISPR's ability to edit genes with unprecedented precision has led some to speculate that military organizations are using this technology to create soldiers with enhanced physical and cognitive abilities.

These theories often cite the Defense Advanced Research Projects Agency (DARPA) as a central figure in these secret programs.DARPA's known investments in genetic research, such as the Safe Genes program, which aims to develop tools for controlling gene editing, add an element of credibility to these claims. The Safe Genes program, launched in 2017, represents a proactive approach to the responsible development and deployment of gene editing technologies. By focusing on safety and control, DARPA aims to mitigate the risks associated with genetic engineering, ensuring that these powerful tools can be used effectively and ethically.

The Safe Genes program has several key objectives. One of its primary goals is to develop technologies that can precisely control when and where gene editing occurs. This includes creating "off-switches" for CRISPR and other gene-editing systems, allowing scientists to halt the editing process if unintended changes are detected. Such control mechanisms are crucial for preventing off-target effects, where edits are made to unintended parts of the genome, potentially causing harmful mutations.

Another important aspect of the Safe Genes program is the development of gene drive technologies. Gene drives are genetic systems that increase the likelihood of a particular gene being passed on to the next generation, thereby spreading genetic modifications through populations more rapidly. While gene drives have significant potential for controlling disease vectors like mosquitoes, they also pose ecological risks if not carefully managed. Safe Genes aims to create reversible gene drives, providing a way to undo genetic modifications if they have unintended consequences.

Furthermore, the program seeks to enhance biosurveillance capabilities to monitor gene editing activities and detect potential misuse. This involves creating diagnostic tools to identify and track the presence of specific genetic modifications in organisms. By improving the ability to monitor genetic changes, DARPA aims to prevent the malicious use of gene editing technologies, such as the creation of genetically modified organisms for bioterrorism.

DARPA's investment in Safe Genes underscores the agency's commitment to advancing genetic research while prioritizing safety and ethical considerations. By developing robust control mechanisms and monitoring tools, DARPA is laying the groundwork for the responsible use of gene editing technologies in both military and civilian contexts. This proactive approach not only enhances the credibility of DARPA's genetic research efforts but also sets a precedent for other organizations engaged in genetic engineering.

The Safe Genes program exemplifies how cutting-edge research can be balanced with ethical responsibility, ensuring that the benefits of gene editing are realized without compromising safety or integrity.

Fiction popularized super soldiers, but recent technological advancements suggest that they may not be as fictional as they seem. In the Captain America series, the super soldier serum made the world safer and better. However, in Marvel's Iron Man 3, a terrorist organization stole the genetic mix to create super soldiers who could survive injuries, regenerate limbs, and demonstrate extraordinary strength, endurance, and agility. Although both the series and the film were fictional, defense applications of emerging genetic technology call that classification into question.

In the US, the Pentagon earmarks considerable resources for human enhancement research that could create enhanced soldiers. The Defense Advanced Research Projects Agency (DARPA) is tasked with this research. In 1958, the US established DARPA in response to the surprise Sputnik launch. Mainly, through programs such as DARPA, the US sought to prevent strategic surprises from negatively impacting its national security. It also aimed to maintain the technological superiority of its military. DARPA is considered the primary innovation engine of the Department of Defense. It uses applied research to address emerging and potential problems. DARPA's six offices include the Biological Technologies Office, Defense Sciences Office, Information Innovation Office,

Microsystems Technology Office, Strategic Technology Office, and Tactical Technology Office.

DARPA recently launched its Biological Technologies Office. In 2016-2017, this office, with its budget of $296 million, explored challenges at the intersection of biology and engineering. DARPA lists several programs focused on self-healing and preventing injuries among soldiers. DARPA's Safe Genes platform specifically protects military personnel from accidental or intentional misuse of genome editing technologies. It states: "Overall, the Safe Genes program is creating a layered, modular, and adaptable solution set to protect warfighters and the homeland against intentional or accidental misuse of genome editing technologies; prevent and/or reverse unwanted genetic changes in a given biological system; and facilitate the development of safe, precise, and effective medical treatments that use gene editors."

With regard to CRISPR technologies in particular, DARPA states: "A University of California, Berkeley team led by Dr. Jennifer Doudna will investigate the development of novel, safe gene editing tools for use as antiviral agents in animal models, targeting the Zika and Ebola viruses. The team will also aim to identify anti-CRISPR proteins capable of inhibiting unwanted genome-editing activity, while developing novel strategies for delivery of genome editors and inhibitors."

It appears that DARPA, and by extension, authorities within the US government, take the threat of unwanted genome editing and genetic modification seriously. But at the same time, DARPA's therapeutic arm, even within the Safe Genes program, self-admittedly seeks to "facilitate the development of safe, precise, and effective medical treatments that use gene editors." Plausibly, as the distinction between therapeutic use and enhancement blurs and alters, concerns arise about enhancement technologies.

Peter Singer of the Brookings Institute reported on DARPA's Metabolically Dominant Soldier program. Writing about DARPA director Callaghan's talk at its 50th anniversary, Singer noted that the US military is studying ways to use "technology and biology to meld man and machine in order to transcend limits of the human body." The project director was quoted as saying, "My measure of success is that the International Olympic Committee bans everything we do."

DARPA has expanded its portfolio to include ambitious projects aimed at writing the Human Genome, pushing the boundaries of genetic science even further. This effort builds on the foundation laid by the Human Genome Project, which was an international scientific research initiative completed in 2003. The Human Genome Project successfully mapped and sequenced the entire human genome, identifying and mapping all the genes of the human species. This monumental achievement provided a comprehensive reference of human DNA and has since revolutionized the fields of medicine, genetics, and biotechnology.

Building on this groundwork, DARPA's current initiatives focus on not just reading, but writing the human genome. This involves synthesizing and assembling entire genomes from scratch, a process that holds the potential for groundbreaking advancements in biotechnology and medicine. One such project is led by the Center of Excellence for Engineering Biology, where scientists are working on the Genome Project-write (GP-write). This initiative aims to develop new technologies for synthesizing large segments of DNA and integrating them into cells, ultimately enabling the creation of custom-designed genomes.

For some scientists at the Center of Excellence for Engineering Biology, the next step involved writing entire genomes and synthesizing them from scratch. DARPA funded the center's Boeke and Harris Wang from Columbia University with $500,000 for a genome writing pilot project. They will use DARPA's funds to engineer human cells that are self-sufficient nutrient factories. By exploiting genes from bacteria, plants, and fungi, this project aims to engineer human cells capable of manufacturing nutrients that un-engineered human cells cannot. In its proposal for synthesizing a prototrophic human genome, the pilot project team noted uses mainly related to combating malnutrition, food shortages, and more economical biosynthesis of medicines. But DARPA's involvement suggested that it sought to use this technology to create self-sustaining soldiers with limited need to eat.

Furthermore, the ability to write genomes opens up possibilities for developing enhanced biological traits that could benefit military personnel. For example, soldiers could be engineered to have increased resistance to diseases, improved physical endurance, or enhanced cognitive abilities. This kind of genetic modification could significantly improve the resilience and effectiveness of military forces, providing a strategic advantage in various operational contexts.

However, these advancements also raise significant ethical, legal, and social questions. The potential for misuse of genome writing technology is a major concern, particularly if it were to be used for creating genetically modified organisms for bioterrorism or other malicious purposes.

To address these concerns, DARPA's Safe Genes program plays a crucial role. Launched to ensure the safe and responsible development of gene-editing technologies, Safe Genes aims to create control mechanisms that can prevent, reverse, or mitigate unintended genetic changes. This includes developing "off-switches" for gene editing systems and strategies for limiting the spread of gene drives—genetic elements that can propagate specific traits through populations.

DARPA's funding of projects aimed at writing the human genome represents a significant leap forward in genetic engineering. While the potential benefits for medical and military applications are immense, they are accompanied by complex ethical and security challenges that must be carefully managed. The legacy of the Human Genome Project provides a strong foundation, but the future of genome writing will require rigorous oversight and international cooperation to ensure that these powerful technologies are used responsibly and ethically.

In recent decades, the military's fascination with enhancing human capabilities has expanded into the realm of genetic technology. Historically, efforts to improve soldier performance included rigorous training programs, performance-enhancing drugs, and advanced prosthetics. However, genetic engineering represents a significant leap forward, promising more profound and permanent changes.

The allure of genetic modification lies in its potential to create super-soldiers with innate advantages that surpass traditional methods of enhancement. Unlike external tools and temporary measures, genetic engineering can alter the very blueprint of human biology, embedding enhancements directly into a soldier's DNA. This shift from external to internal modification reflects a broader trend in military science: the pursuit of permanent, reliable, and heritable traits that can be passed down through generations. This transition also signifies a move towards a more holistic approach to soldier enhancement, where the focus is not just on physical capabilities but also on cognitive and psychological resilience.

Consider the ability to engineer blood to allow soldiers to breathe underwater or remain unaffected by altitude sickness. Such advancements would render traditional geographical barriers—natural defenses that many nations have relied upon for self-protection—ineffective. In the high altitudes of the Himalayas, for instance, tensions continue to rise among neighboring countries. Enhanced soldiers equipped with respirocytes, or artificial red blood cells, could operate in these low-oxygen environments as efficiently as they would at sea level. This technological breakthrough would provide a significant tactical advantage in mountainous regions where thin air typically hampers human performance.

Respirocytes are designed to vastly outperform natural red blood cells in their ability to carry oxygen and remove carbon dioxide. These artificial cells can be engineered to store and release gasses at controlled rates, ensuring that soldiers maintain optimal physiological function even in the most challenging conditions. This means that troops could conduct prolonged operations in underwater environments or at high altitudes without the need for acclimatization or supplemental oxygen.

The strategic implications of such advancements are profound. Enhanced soldiers would not only overcome the physical limitations imposed by diverse environments but would also gain a formidable edge over adversaries. Nations lacking the resources to invest in genetic and biotechnological enhancements, or those bound by ethical and moral constraints against such modifications, would find themselves at a distinct disadvantage. This could lead to a significant shift in the balance of power, with technologically advanced militaries able to project force and conduct operations in previously inaccessible or inhospitable areas.

Furthermore, the ability to operate seamlessly across different terrains would enable enhanced soldiers to undertake covert missions and surprise maneuvers, exploiting the element of unpredictability. For instance, troops could infiltrate enemy lines via underwater routes or establish high-altitude bases that are difficult to detect and attack. This would force opposing forces to rethink their defense strategies and invest in countermeasures, potentially sparking a new arms race focused on biological and genetic enhancements.

Beyond immediate military applications, the development of such technologies could have broader implications for global security and warfare ethics. The use of respirocytes and other enhancements could blur the lines between human and machine, raising questions about the nature of soldiering and the moral responsibilities of military organizations. As these technologies become more sophisticated, the debate over their ethical use will intensify, challenging existing norms and potentially leading to new international regulations and treaties aimed at governing their deployment.

The ability to engineer soldiers capable of thriving in extreme conditions would revolutionize military operations, diminishing the protective value of geographical barriers and giving rise to new strategic possibilities. As nations grapple with the implications of these advancements, the balance of power could shift dramatically, underscoring the importance of ethical considerations and international cooperation in the era of enhanced warfare.

CRISPR technology holds the potential to revolutionize military operations by engineering soldiers with enhanced resilience to environmental extremes, such as extreme cold or heat. By precisely editing specific genes, scientists can potentially create soldiers who can endure and perform optimally in harsh climates, thereby expanding the scope of military operations in regions previously considered too challenging for human deployment.

To understand how CRISPR could make this a reality, let's delve into the specifics. One key area of focus could be the enhancement of thermoregulation, the body's ability to maintain its core temperature. For example, genes that regulate the production of antifreeze proteins found in certain Arctic fish could be introduced into human DNA. These proteins prevent ice crystals from forming in the blood, allowing organisms to survive in freezing temperatures. Similarly, genes responsible for heat shock proteins, which help protect cells from stress caused by extreme heat, could be upregulated. This would enable soldiers to operate in scorching deserts without succumbing to heatstroke or dehydration.

Another potential application of CRISPR is the enhancement of metabolic efficiency. By modifying genes involved in energy metabolism, soldiers could be engineered to have a higher basal metabolic rate, allowing them to generate more internal heat in cold environments. Conversely, in hot climates, the same genetic modifications could help improve the body's ability to dissipate heat more effectively. For instance, the overexpression of genes responsible for vasodilation could enhance blood flow to the skin, promoting heat loss.

The implications of such advancements are profound. Consider military operations in the Arctic, where extreme cold and ice-covered terrain present significant challenges. Currently, soldiers require extensive training and specialized equipment to survive in these conditions. With CRISPR-engineered resilience, soldiers could undertake prolonged missions in the Arctic without the need for bulky cold-weather gear. This would allow for greater mobility, stealth, and endurance, giving a tactical advantage in these environments.

In hot, arid regions like the Middle East or Africa, where temperatures can soar to unbearable levels, CRISPR could similarly enhance operational capabilities. Soldiers who are genetically modified to withstand extreme heat could carry out missions in desert environments without the risk of heat exhaustion. This would not only improve their effectiveness but also reduce the logistical burden of supplying water and cooling systems.

The potential to operate in such diverse and extreme environments would fundamentally change military strategy and planning. Nations with the capability to deploy genetically enhanced soldiers would have a significant advantage in global conflicts. They could project power into regions previously deemed inhospitable, execute surprise operations, and maintain a presence in areas where traditional forces would struggle to survive.

However, the path to realizing these capabilities is fraught with scientific, ethical, and regulatory challenges. The science of gene editing is still in its early stages, and the long-term effects of genetic modifications are not fully understood. There is also the risk of unintended consequences, such as off-target effects that could cause harmful mutations. Moreover, the ethical implications of creating genetically modified soldiers are profound. Questions about consent, the potential for coercion, and the broader impact on society must be carefully considered.

To address these challenges, rigorous testing and oversight would be necessary. International cooperation and the establishment of clear ethical guidelines will be crucial to ensuring that the development and deployment of such technologies are conducted responsibly. Regulatory frameworks would need to be established to govern the use of genetic modifications in the military, balancing the potential benefits with the need to protect human rights and prevent misuse.

While DARPA's ambitious projects in genome writing and genetic modification hold immense potential, there is a significant counterargument to consider: the military's rush to be the first in the world with this technology may lead to unforeseen and potentially disastrous consequences. The intense pressure to achieve a strategic advantage could result in hasty implementation of gene editing techniques without fully understanding the long-term effects or ethical implications.

Firstly, the race to be first often leads to cutting corners in research and development. In the military context, this urgency can mean deploying unproven technologies in high-stakes environments. The complexity of genetic engineering requires thorough testing and evaluation to ensure safety and efficacy. Rushing this process increases the risk of unintended side effects, such as genetic mutations or unforeseen health problems in modified individuals. These risks are not just theoretical; they have been observed in other fields of biotechnology where premature deployment led to significant setbacks and public distrust.

Moreover, the lack of comprehensive ethical frameworks exacerbates the potential for abuse and misuse of genetic technologies. In the scramble to achieve dominance, ethical considerations might be sidelined, leading to actions that could violate human rights and international norms. For instance, the prospect of creating genetically enhanced soldiers raises serious questions about consent, autonomy, and the potential for coercion.

Soldiers might be pressured or even forced to undergo genetic modifications, compromising their personal agency and subjecting them to unknown health risks.

Furthermore, the environmental and ecological risks associated with genetic modifications are not fully understood. The deployment of genetically modified organisms, including humans, into various environments could have unforeseen impacts on ecosystems and biodiversity. These changes could be irreversible, leading to long-term consequences that we are currently ill-equipped to manage.

Finally, the geopolitical ramifications of a genetic arms race cannot be ignored. As countries vie for supremacy in genetic technologies, the likelihood of international tensions and conflicts increases. This competition could lead to a destabilizing arms race, with nations prioritizing genetic enhancements over diplomatic solutions and cooperative security measures. The history of nuclear proliferation offers a sobering parallel, illustrating how the pursuit of technological dominance can escalate into global crises.

While the potential benefits of genetic technologies are substantial, the rush to be the first in the world to deploy them carries significant risks. The military's urgency to achieve a strategic advantage must be balanced with a cautious and ethical approach to research and development. Ensuring rigorous testing, comprehensive ethical oversight, and international cooperation is essential to harness the benefits of genetic engineering while mitigating its potential dangers.

CRISPR technology offers the potential to engineer soldiers with enhanced resilience to environmental extremes, significantly expanding the scope of military operations in harsh climates. While the scientific and ethical hurdles are considerable, the strategic advantages of such capabilities make it a compelling area of research. As the technology advances, it will be essential to navigate these challenges thoughtfully to harness the benefits while mitigating the risks.

Let's take a look at another possibility in the realm of CRISPR soldiers. The technology could be utilized to develop advanced bio-sensors, integrating soldiers' biological systems with cutting-edge genetic modifications to detect and respond to chemical or biological agents more effectively. Imagine a future where soldiers possess a built-in warning system, an innate ability to sense harmful substances in their environment before they can cause harm.

These genetically engineered bio-sensors could revolutionize the detection of chemical warfare agents by significantly enhancing the natural sensory capabilities of human cells. For example, olfactory receptors, which are responsible for our sense of smell, could be genetically modified to become hypersensitive to specific hazardous chemicals, such as sarin or mustard gas. This enhancement would involve tweaking the receptors' genetic code to increase their sensitivity and specificity to these toxic substances.

When these augmented receptors detect even trace amounts of chemical agents, they could initiate a rapid biological response. This detection could trigger an immediate alert within the soldier's body, functioning as an internal alarm system. Such an alert might involve the release of specific biomarkers into the bloodstream, which could be detected by wearable health monitors. These monitors, integrated with advanced AI algorithms, could then analyze the biomarker data in real-time and send an instant alert to both the soldier and their command center, ensuring a swift and coordinated response to the threat.

Moreover, these bio-sensors could be integrated with other physiological monitoring systems to provide a comprehensive picture of the soldier's environment and health status. For instance, they could work in tandem with sensors that measure vital signs such as heart rate, respiration, and body temperature, offering a holistic view of the soldier's readiness and exposure to hazardous conditions.

Beyond merely detecting harmful agents, these bio-sensors could initiate a biological response to neutralize the threat, fundamentally transforming the body's defensive capabilities. Imagine a scenario where a soldier, upon exposure to a toxic chemical, experiences an automatic activation of genetic modifications designed to counteract the threat. This could involve the engineered production of specific proteins or enzymes that rapidly break down the harmful compound into non-toxic components, effectively neutralizing the danger.

For instance, consider a soldier who inhales a toxic nerve agent. Traditional protective measures might include immediate evacuation and administration of antidotes, which can be time-consuming and may not be feasible in the heat of battle. With advanced bio-sensors, the soldier's body could detect the presence of the nerve agent almost instantaneously. This detection would trigger a cascade of genetic responses, activating genes responsible for producing enzymes that degrade the nerve agent into harmless molecules. The entire process could occur within seconds to minutes, significantly reducing the agent's harmful effects.

This rapid, automatic response would not only protect the individual soldier but also serve a broader protective function. By neutralizing the toxic substance at the point of entry, the engineered biological response would prevent the spread of the agent to others, reducing the risk of secondary exposure to fellow soldiers and civilians. This could be particularly crucial in confined spaces or populated areas, where the dissemination of harmful agents poses a significant threat.

Moreover, the implications of such technology extend beyond chemical threats. Bio-sensors could be programmed to detect and respond to a range of biological hazards, including pathogenic bacteria and viruses. Upon identification of an infectious agent, the soldier's body could initiate the production of antiviral proteins or antibacterial peptides, effectively combating the infection before it takes hold. This preemptive strike against pathogens could drastically reduce infection rates and improve survival outcomes in the field.

The integration of bio-sensors and engineered biological responses exemplifies the convergence of biotechnology and defense. It highlights a future where soldiers are not just protected by external gear but are inherently equipped with enhanced biological defenses. This innovation represents a paradigm shift in military strategy, where the human body itself becomes a sophisticated platform for defense and resilience, capable of adapting to and neutralizing a wide array of threats in real-time. The potential benefits of such advancements underscore the importance of continued research and development in this area, promising a new era of soldier protection and battlefield safety.

The potential applications extend beyond chemical and biological detection. Imagine a future where soldiers possess vision capabilities far beyond what is currently possible. Through the application of CRISPR technology, visual receptors in the eyes could be genetically enhanced to detect a broader spectrum of light, including ultraviolet (UV) and infrared (IR). This enhancement would allow soldiers to operate in conditions that are typically challenging, such as complete darkness, dense smoke, or even through environmental obstructions.

Enhanced UV vision in CRISPR soldiers would be possible through the genetic modification of photoreceptor cells in the retina, which are responsible for detecting light. The human retina contains two main types of photoreceptor cells: rods and cones. Rods are responsible for vision in low light conditions but do not detect color, while cones are responsible for color vision and operate best in bright light. Humans typically have three types of cones, each sensitive to different wavelengths of light corresponding to blue, green, and red.

To enable UV vision, CRISPR could be used to introduce or modify genes that produce photopigments sensitive to UV light. Scientists would start by identifying and isolating genes from organisms that naturally possess UV vision. Many birds, insects, and some mammals have photopigments that are sensitive to UV light. For example, certain bird species have a fourth type of cone cell that allows them to see UV light. Using CRISPR-Cas9, these UV-sensitive genes could be inserted into the DNA of human photoreceptor cells. This involves creating a guide RNA (gRNA) that matches the target site in the human genome where the UV-sensitive gene will be inserted. The Cas9 enzyme, guided by the gRNA, makes a precise cut at the target site. A DNA template containing the UV-sensitive gene is then introduced, and the cell's natural repair mechanisms incorporate the new gene into the genome through a process called homology-directed repair (HDR).

The inserted gene must be properly expressed in the photoreceptor cells to produce the UV-sensitive photopigment. Scientists would ensure that the new gene integrates seamlessly into the existing genetic framework of the cones, enabling the production of the UV-sensitive protein. After genetic modification, the functionality of the newly introduced UV-sensitive cones would be tested. This could involve both in vitro (in the lab) and in vivo (in living organisms) experiments to ensure that the modified cells respond to UV light as intended. Animal models, such as genetically modified mice, are often used in initial experiments to validate the functionality and safety of the new photoreceptor cells before considering human applications.

UV vision could help soldiers detect hidden threats or hazards that are not visible in normal light conditions. For instance, many substances fluoresce under UV light, which could be used to detect chemical or biological agents. UV vision could complement other visual enhancements like infrared, allowing soldiers to see in a wider range of lighting conditions without relying on external equipment. However, there are significant ethical and safety concerns. Genetic modifications, especially those involving human vision, raise issues about consent, potential long-term effects, and the implications of creating genetically enhanced humans. Integrating new photoreceptors into the complex architecture of the human eye is technically challenging. Ensuring that these modifications do not interfere with existing vision and that they function correctly in the highly specialized environment of the retina is a major hurdle. Any genetic modification intended for use in humans would need to pass stringent regulatory hurdles to ensure safety and efficacy. This involves extensive preclinical and clinical testing.

By leveraging CRISPR technology, scientists could theoretically introduce UV sensitivity into human vision, potentially creating soldiers with enhanced visual capabilities. However, the process involves complex genetic engineering and thorough validation to ensure it is safe and effective.

Infrared vision, on the other hand, could grant soldiers the extraordinary ability to perceive thermal signatures. This advanced capability would enable them to detect hidden enemies by their body heat, even in total darkness or through dense foliage and smoke, effectively rendering traditional methods of concealment and camouflage obsolete. By visualizing the heat emitted by living beings, soldiers equipped with infrared vision could identify threats that would otherwise remain unseen, providing a significant tactical advantage in combat situations.

The applications of thermal vision extend far beyond mere enemy detection. In disaster zones, soldiers with infrared vision could locate survivors trapped under rubble or debris by detecting their body heat, significantly improving the efficiency and speed of rescue operations. This technology could also prove invaluable in search and rescue missions in challenging environments, such as dense forests, mountainous terrains, or collapsed structures, where conventional visual aids fail.

Moreover, infrared vision could be employed to monitor and identify overheating machinery or vehicles, preventing potential mechanical failures or malfunctions before they occur. This proactive approach to maintenance could enhance the reliability and longevity of military equipment, reducing the risk of sudden breakdowns during critical missions.

In various operational environments, enhanced situational awareness provided by thermal vision could be a game-changer. Soldiers could navigate through smoke-filled battlefields or areas with low visibility, maintaining their orientation and identifying potential hazards with ease. The ability to see through environmental obstructions would also allow for more effective planning and execution of tactical maneuvers, giving military forces a distinct edge in both offensive and defensive operations.

Infrared vision could further enhance coordination and communication within military units. By enabling soldiers to see each other's thermal signatures, even in complete darkness, unit cohesion and synchronization during night operations would be significantly improved. This could lead to more effective team strategies and reduced risks of friendly fire incidents.

Beyond the battlefield, the integration of thermal vision into civilian applications holds promise as well. Law enforcement agencies could use this technology to track suspects in urban environments or monitor large crowds for unusual activity. Firefighters could benefit from infrared vision to navigate through smoke-filled buildings and locate individuals in need of rescue, while industrial workers could use it to detect overheating equipment and prevent potential hazards.

The implementation of infrared vision technology offers a multitude of benefits across various fields. For military personnel, it provides enhanced detection capabilities, improved rescue operations, better equipment maintenance, and superior situational awareness. Its potential applications in civilian sectors further underscore the transformative impact of this technology on safety, efficiency, and overall operational effectiveness.

Moreover, the integration of UV and IR vision could improve target acquisition and identification, allowing soldiers to distinguish between friend and foe more effectively in chaotic and low-visibility conditions. This multi-spectral vision could also aid in navigation and coordination during night operations, reducing reliance on external night-vision devices and enhancing overall combat effectiveness.

Similarly, auditory receptors could be genetically modified to detect a broader spectrum of sound frequencies, significantly enhancing a soldier's ability to perceive distant or subtle noises that might indicate enemy presence. By expanding the range of detectable frequencies, soldiers could hear high-frequency sounds emitted by electronic devices, or low-frequency vibrations caused by movement, that would otherwise be imperceptible to the human ear.

This enhancement could provide a tactical advantage in various combat scenarios. For example, soldiers with enhanced auditory capabilities might be able to hear the faint hum of a distant drone, the rustling of leaves that betrays an enemy's stealthy approach, or even the whispered communications of adversaries in the next building. By distinguishing these sounds from the background noise of the battlefield, enhanced soldiers could respond more quickly to threats, setting up ambushes, avoiding detection, or coordinating their movements more effectively.

Moreover, enhanced auditory receptors could be fine-tuned to filter out specific frequencies associated with common battlefield noise, such as gunfire or explosions, reducing the risk of auditory overload and allowing soldiers to maintain focus during intense combat situations. This genetic modification could also be integrated with advanced communication systems, offering a futuristic leap in battlefield communication technology. By incorporating specific genes that enhance neural interface capabilities, soldiers could have direct, secure connections to sophisticated communication networks. This would allow them to receive and process encrypted audio messages directly through their nervous system, bypassing the need for traditional devices like radios or headsets.

Such a system would enable soldiers to receive real-time, encrypted instructions and intelligence updates that are completely undetectable by conventional surveillance and interception methods. The genetic enhancements could include bioengineered neural receptors tuned to specific frequencies, effectively making the soldier's brain a living, encrypted communication hub. This would ensure that sensitive information remains secure even in the most hostile and surveillance-intensive environments.

Additionally, these enhancements could enable seamless communication with other enhanced soldiers and command centers, creating a networked battlefield where every operative is connected through a highly secure, biologically integrated communication system. This integration would not only improve coordination and response times but also reduce the cognitive load on soldiers by delivering critical information in a more intuitive and less obtrusive manner. The potential for such technology underscores the military's interest in blending biological and technological advancements to create a new generation of highly capable and connected soldiers.

Research in this area delves into the auditory capabilities of animals known for their exceptional hearing, with the aim of uncovering the genetic foundations of these abilities and potentially transferring similar traits to humans. For instance, bats possess an extraordinary capacity for echolocation, allowing them to navigate and hunt in complete darkness using ultrasonic frequencies. This sophisticated biological sonar system is rooted in their specialized auditory genes and neurological pathways. Echolocation in bats involves the emission of high-frequency sound waves that bounce off objects and return as echoes, providing a detailed acoustic map of their surroundings.

To achieve this remarkable feat, bats have evolved a suite of specialized genes and highly sensitive auditory systems. These adaptations enable them to produce and detect ultrasonic sounds that are far beyond the range of human hearing. At the genetic level, bats possess unique variants of genes involved in hearing and sound production. For example, genes that encode for proteins in the inner ear, such as prestin, are fine-tuned to enhance sensitivity to high-frequency sounds. Prestin, a motor protein in the outer hair cells of the cochlea, plays a crucial role in amplifying sound vibrations, allowing bats to detect even the faintest echoes.

Neurologically, bats have highly developed auditory cortices that process these ultrasonic frequencies with remarkable precision. The neural pathways in a bat's brain are specifically adapted to interpret the rapid and complex echoes received during flight. This involves a sophisticated system of timing and frequency analysis that enables bats to discern the size, shape, distance, and even texture of objects in their environment.

By studying the genetic and physiological mechanisms that enable bats to process high-frequency sounds, scientists can identify key genes and proteins responsible for their acute hearing. Advanced techniques in genomics and bioinformatics allow researchers to sequence and analyze the bat genome, pinpointing genetic variations that contribute to their echolocation abilities. Functional studies, such as gene editing and expression analysis, further elucidate how these genetic factors translate into the physiological traits observed in bats.

This research has broader implications beyond understanding bat biology. Insights gained from studying bat echolocation can inform the development of advanced technologies in various fields. For example, bioinspired sonar systems and acoustic sensors could improve navigation and detection capabilities in autonomous vehicles and robotics. Additionally, understanding the genetic basis of echolocation may lead to novel treatments for hearing impairments in humans, as it provides a blueprint for enhancing auditory function through genetic and molecular approaches.

The study of bat echolocation exemplifies how examining the natural world can yield valuable insights into genetics and physiology, with far-reaching applications in technology and medicine. Through interdisciplinary research, scientists can harness the power of nature's innovations to address complex challenges and improve human life.

Similarly, elephants are remarkable for their ability to communicate through infrasound, which consists of sound waves at frequencies below the range of human hearing. These low-frequency sounds can travel long distances, allowing elephants to communicate over several kilometers. The genetic basis for this unique capability lies in the structure and function of their vocal cords and inner ear. By examining the genetic blueprints that allow elephants to produce and perceive infrasound, researchers can gain insights into the adaptations that make this communication possible.

To harness these abilities for human enhancement, scientists could use techniques like CRISPR to edit the human genome, introducing specific genes that confer enhanced auditory capabilities. For example, by incorporating genes responsible for the sensitivity to ultrasonic frequencies found in bats, humans might develop the ability to perceive a broader range of sounds, improving their situational awareness and communication in environments where conventional hearing is inadequate. Similarly, integrating genes that allow elephants to detect infrasound could enhance human capabilities in monitoring seismic activities or improving long-distance communication.

Beyond theoretical applications, the practical implications of such genetic enhancements are vast. Enhanced auditory senses could benefit military personnel by improving their ability to detect threats, communicate covertly, or navigate complex environments.

In civilian contexts, individuals with superior hearing might excel in fields such as music, surveillance, or search and rescue operations. Additionally, these advancements could aid people with hearing impairments, offering new avenues for treatment and rehabilitation.

However, this line of research also raises significant ethical and safety concerns. The long-term effects of introducing foreign genes into the human genome are still unknown, and the potential for unintended consequences must be carefully considered. Ethical debates surrounding genetic modification, particularly in enhancing human abilities beyond natural limits, necessitate rigorous scrutiny and robust regulatory frameworks to ensure responsible development and application of these technologies. By balancing scientific innovation with ethical responsibility, researchers can explore the potential of genetic enhancements while safeguarding human dignity and well-being.

The implications of such advancements are profound, raising questions about the ethical use of genetic modifications in warfare and the potential long-term effects on soldiers' health and well-being. Nonetheless, the pursuit of these enhancements underscores the ongoing quest to push the boundaries of human performance on the battlefield, leveraging cutting-edge science to gain a strategic edge over adversaries.

These advancements in bio-sensor technology could greatly enhance situational awareness and survivability on the battlefield. By providing soldiers with real-time, precise information about their surroundings and potential threats, these enhancements could lead to more informed decision-making and quicker reactions to dangers. This convergence of biology and technology epitomizes the next frontier in military innovation, aiming to create soldiers who are not only physically superior but also equipped with enhanced perceptual abilities that far surpass those of unmodified humans.

The potential applications of these bio-sensors are vast and varied. In addition to protecting soldiers on the battlefield, they could be used in civilian settings to safeguard against bioterrorism or industrial accidents involving hazardous materials. By continually monitoring their environment and responding to threats in real-time, individuals equipped with these bio-sensors would embody a new level of biological preparedness and resilience.

Developing such technology requires a multidisciplinary approach, combining advances in genetic engineering, synthetic biology, and bioinformatics. Researchers would need to identify the key genes and pathways involved in sensory detection and response, then engineer these components to function in a highly specific and controlled manner. This would involve extensive testing and optimization to ensure that the bio-sensors are both effective and safe.

Ethical considerations must also be addressed, particularly regarding the potential for misuse or unintended consequences. The idea of altering human biology to create super-soldiers with enhanced sensory capabilities raises important questions about consent, equity, and the long-term impacts on human health and society.

The utilization of CRISPR and other genetic technologies to develop bio-sensors represents a revolutionary step forward in military and civilian defense. By equipping individuals with the ability to detect and respond to chemical and biological threats on a cellular level, we could vastly improve our preparedness and resilience in the face of emerging dangers. However, careful consideration and regulation will be crucial to ensuring that these advancements are used responsibly and ethically.

This need for responsible use is underscored by historical and contemporary conspiracy theories that highlight public fears about the misuse of advanced technologies. One of the most infamous conspiracy theories is the Montauk Project, which alleges that the U.S. government conducted clandestine experiments at Camp Hero, a decommissioned military base in Montauk, New York. According to proponents of this theory, these experiments were part of a covert program aimed at pushing the boundaries of scientific and military capabilities. The alleged activities at Camp Hero included mind control, time travel, and genetic modification of humans to create super-soldiers.

Supporters of the Montauk Project claim that these experiments were conducted on unwitting participants, including children, and involved highly advanced technology that remains undisclosed to the public. Stories of bizarre occurrences, such as individuals reportedly traveling through time or exhibiting extraordinary mental abilities, are central to the lore of the Montauk Project. These tales often intertwine with other conspiracy theories, suggesting a web of secret government programs all aimed at developing superhuman capabilities and ultimate military superiority.

Despite the lack of credible evidence to substantiate these extraordinary claims, the story of the Montauk Project has persisted in popular culture. It has inspired numerous books and documentaries that explore the supposed activities and their implications. One of the most notable cultural impacts of the Montauk Project conspiracy theory is its influence on the hit television series "Stranger Things." The show's premise of secret government experiments, telekinetic children, and parallel dimensions mirrors the alleged activities at Camp Hero, bringing the Montauk Project into the mainstream consciousness.

The enduring fascination with the Montauk Project reflects a broader societal intrigue with the idea of hidden government agendas and the potential for human enhancement. Whether through genetic modification, advanced technology, or other means, the concept of creating super-soldiers taps into deep-seated fears and aspirations about the future of humanity. As with many conspiracy theories, the Montauk Project continues to captivate the imagination, blurring the line between fiction and reality and fueling ongoing debates about the limits of scientific and military experimentation.

The persistence of these conspiracy theories highlights significant ethical and legal concerns regarding genetic research. The notion that governments might engage in clandestine genetic experimentation raises questions about oversight, consent, and the potential for abuse. In 2015, a group of leading scientists and ethicists, including CRISPR pioneers Jennifer Doudna and Emmanuelle Charpentier, called for a global moratorium on human germline editing until ethical and safety concerns could be addressed. This call underscores the broader apprehensions within the scientific community about the potential for misuse of genetic technologies.

The media plays a crucial role in shaping public perception of genetic engineering and military applications. Sensationalized reports and fictional portrayals of super-soldiers contribute to the proliferation of conspiracy theories. Movies like "Universal Soldier" and video games like "Metal Gear Solid" depict genetically enhanced warriors, blurring the lines between science fiction and potential reality. These portrayals can fuel public fear and fascination, creating an environment where conspiracy theories thrive.

While it is essential to approach these conspiracy theories with skepticism, it is equally important to consider the realistic possibilities of genetic enhancements in the military. The concept of super-soldiers is not entirely far-fetched; the U.S. military has openly explored various enhancement technologies, including exoskeletons and cognitive enhancers. The DARPA-funded "Brain-Machine Interface" program, for instance, aims to develop devices that enable direct communication between the brain and external devices, potentially enhancing soldier capabilities.

However, the leap from current research to the creation of CRISPR-modified super-soldiers involves numerous scientific, ethical, and logistical challenges. The complexities of human genetics, combined with the potential for unintended consequences, make the scenario of fully optimized soldiers a distant possibility rather than an imminent reality.

Conspiracy theories about military experiments and super-soldiers, though largely speculative, reflect genuine concerns about the potential misuse of genetic technologies. These theories serve as a cautionary tale, emphasizing the need for transparency, ethical oversight, and public dialogue as we navigate the uncharted waters of genetic engineering in military contexts. As we move forward, it is crucial to balance the pursuit of scientific advancements with the safeguarding of human rights and ethical principles.

In the following chapter, we will delve deeper into the specific enhancements that CRISPR technology could potentially offer to soldiers, exploring both the scientific breakthroughs and the ongoing research that could bring these theories closer to reality.

Chapter 3.

Enhancing the Human Soldier

Continuing from the previous discussion on the historical context and modern military interests in genetic technology, it's clear that the potential applications of CRISPR in the military sphere are vast and unprecedented. The idea of genetically enhancing soldiers raises numerous possibilities, each with its own set of implications. As we delve into these potential enhancements, it's important to examine not only the scientific feasibility but also the profound impact these advancements could have on warfare and society.

One of the most straightforward applications of CRISPR in enhancing soldiers lies in its potential to significantly increase physical strength and endurance. Advances in genetic research have unveiled key insights into the genetic basis of muscle growth and repair, providing a promising avenue for military applications. Central to this research is the MSTN gene, which encodes myostatin, a protein known for its role in inhibiting muscle growth.

Myostatin acts as a natural brake on muscle development, ensuring that muscle growth does not occur unchecked. However, in the context of enhancing soldier performance, reducing or eliminating the activity of myostatin could yield substantial benefits.

Researchers have already achieved remarkable results in animal studies by inhibiting myostatin, a protein that inhibits muscle growth. These studies have demonstrated that animals with reduced myostatin levels exhibit significantly increased muscle mass and strength. For instance, myostatin-null mice, often referred to as "mighty mice," develop muscle mass far beyond that of their normal counterparts, boasting a physique that is significantly more muscular and stronger. These "mighty mice" not only display larger muscle size but also enhanced physical capabilities, making them a prime example of the potential for genetic modifications to enhance physical attributes.

Similar results have been observed in larger animals, such as cattle and dogs. In cattle, the inhibition of myostatin leads to a condition known as "double muscling," where the animals develop exceptionally large and well-defined muscles. This trait is highly valued in the agricultural industry for producing leaner meat with greater efficiency. Belgian Blue cattle, a breed known for its myostatin mutation, exhibit this pronounced muscle hypertrophy, resulting in a significant increase in muscle mass compared to standard cattle breeds. The enhanced physical performance and increased muscle growth in these cattle illustrate the profound impact that myostatin inhibition can have on an organism's physical characteristics.

In dogs, particularly in breeds such as the Whippet, a mutation that reduces myostatin activity has led to the development of exceptionally muscular and agile animals. These dogs, often referred to as "bully Whippets," exhibit greater muscle mass and strength, which translates to superior athletic performance. The enhanced muscle growth observed in these dogs demonstrates the potential for genetic modifications to significantly alter and improve physical abilities in mammals.

The overbreeding of dogs, driven by the pursuit of specific traits and physical appearances, has led to a host of genetic illnesses and diseases. This issue offers a cautionary parallel to the potential pitfalls of the overuse of genetic editing in humans. Both scenarios underscore the risks of prioritizing desired traits without fully understanding or considering the broader genetic implications.

In the case of dogs, selective breeding practices have often focused on aesthetic traits, such as coat color, size, and body shape. This narrow focus has led to a reduction in genetic diversity and the amplification of harmful recessive genes. For example, many purebred dogs suffer from inherited conditions like hip dysplasia, heart defects, and respiratory issues. Bulldogs, with their distinctively flat faces, are prone to brachycephalic airway syndrome, which causes severe breathing difficulties. Similarly, German Shepherds are frequently affected by hip dysplasia due to their exaggerated hind leg angulation, which was selected for appearance rather than function.

These health issues in dogs highlight the dangers of genetic manipulation without comprehensive foresight. The same principles apply to the potential overuse of genetic editing in humans. CRISPR and other gene-editing technologies offer the promise of eliminating genetic diseases and enhancing human traits. However, the rush to apply these technologies could lead to unforeseen consequences, mirroring the pitfalls seen in dog breeding.

One major risk is the reduction of genetic diversity. Just as dog breeds have become genetically uniform, widespread use of genetic editing to select for specific human traits could reduce the genetic variability that is crucial for the resilience of our species. This lack of diversity could make humans more susceptible to new diseases and environmental changes, as a narrower gene pool means fewer adaptive traits.

Moreover, the pursuit of "designer" traits in humans could inadvertently amplify harmful genetic mutations. The complexity of the human genome means that editing one gene can have cascading effects on other genes and biological systems. This interconnectedness could lead to the emergence of new health issues that are currently unforeseen. For instance, editing genes to enhance physical strength or cognitive abilities might unintentionally disrupt other vital functions, leading to new forms of genetic disorders.

The ethical implications of genetic editing also mirror the concerns in dog breeding. Selective breeding in dogs has often been driven by superficial and arbitrary standards, sometimes at the expense of the animal's well-being. In humans, the pressure to conform to societal ideals of beauty, intelligence, or athleticism could lead to ethical dilemmas about what constitutes "desirable" traits. This could exacerbate social inequalities and create new forms of discrimination against those who are not genetically enhanced.

Furthermore, the psychological and social impacts of genetic editing must be considered. Just as purebred dogs may suffer from behavioral issues due to their genetic background, humans could face identity and mental health challenges if genetic editing is used to enforce certain traits. The knowledge that one's abilities and characteristics were artificially selected could impact self-esteem and societal dynamics, leading to a range of psychological effects.

The overbreeding of dogs and the potential overuse of genetic editing in humans share significant parallels. Both highlight the dangers of manipulating genetics without fully understanding the long-term consequences. The health problems seen in purebred dogs serve as a stark reminder of the importance of maintaining genetic diversity and exercising caution in genetic interventions. As we stand on the brink of a new era in genetic engineering, it is crucial to learn from these lessons and approach human genetic editing with a balanced and ethical perspective, ensuring that the benefits do not come at the cost of unforeseen genetic and societal pitfalls.

Despite these warnings, the success of animal studies in genetic modification has paved the way for exploring similar interventions in humans. By understanding and manipulating the pathways that regulate muscle growth, researchers aim to develop treatments for muscle-wasting diseases, such as muscular dystrophy, and potentially enhance human physical capabilities. The promising results from animal models provide a strong foundation for future research, highlighting the potential for myostatin inhibition to revolutionize the field of genetic engineering and human enhancement. The implications of these findings extend beyond medical applications, hinting at a future where enhanced physical performance through genetic modification could become a reality.

Moreover, the ability to enhance muscle repair mechanisms through genetic modifications could significantly augment soldier endurance and resilience. In addition to myostatin inhibition—a well-known method for promoting muscle growth—researchers are exploring other genetic pathways involved in muscle regeneration and repair. One such promising area of study is the activation of satellite cells, which are a type of stem cell found within muscle tissue. These cells play a crucial role in muscle tissue recovery, as they are responsible for repairing and rebuilding damaged muscle fibers.

Enhancing the activity of satellite cells through genetic modifications could revolutionize the way soldiers recover from injuries. By boosting the regenerative capacity of these cells, it is possible to accelerate the healing process, allowing soldiers to recover from wounds and physical stress more rapidly. This would not only reduce the downtime associated with injuries but also enhance overall mission readiness and effectiveness. Soldiers could maintain peak physical condition and quickly bounce back from the strains of combat, ensuring that they remain combat-ready even after sustaining injuries.

Furthermore, genetic enhancements could be meticulously designed to boost the production of specific growth factors and proteins that play crucial roles in muscle repair and regeneration. Among these, insulin-like growth factor 1 (IGF-1) stands out as a prime candidate due to its potent effects on muscle growth and recovery. IGF-1 is a naturally occurring hormone that significantly influences muscle development by stimulating protein synthesis and increasing the proliferation and differentiation of muscle cells. It has been extensively studied for its role in enhancing muscle hypertrophy and improving healing processes following injury.

Imagine a scenario where soldiers are genetically engineered to produce elevated levels of IGF-1 in their muscle tissues. This genetic modification could enable their muscles to grow larger and stronger at an accelerated rate compared to unmodified humans. The increased IGF-1 levels would not only enhance muscle mass and strength but also improve the rate at which muscle tissues recover from the stresses of intense physical activity and injury. This means that soldiers could potentially endure more rigorous training regimens and recover more rapidly from the physical toll of combat, reducing downtime and maintaining peak operational readiness.

The implications of such genetic modifications extend beyond merely enhancing muscle performance. Soldiers with enhanced IGF-1 production could also experience improved overall physical resilience. The accelerated repair and regeneration of muscle tissues would contribute to faster recovery from wounds, allowing these genetically modified soldiers to return to the battlefield more quickly after sustaining injuries. This rapid healing capability could be a game-changer in combat situations, where the ability to swiftly recover from physical damage can be the difference between life and death, victory and defeat.

Moreover, the strategic advantages of such enhancements could be profound. Armies equipped with soldiers who possess superior physical capabilities and faster recovery times would have a significant edge in sustained engagements. These enhanced soldiers could carry heavier loads, traverse challenging terrains with greater ease, and maintain high levels of performance under extreme conditions. The ability to recover quickly would also mean that these soldiers could be deployed more frequently and for longer durations, maximizing their utility and effectiveness in various military operations.

By leveraging the power of CRISPR and other gene-editing technologies, researchers are exploring the frontiers of human potential, aiming to create a new breed of warriors with unparalleled physical capabilities. These advancements, while still in the experimental stages, point to a future where genetic modifications could redefine the limits of human performance, pushing the boundaries of what soldiers can achieve on the battlefield.

Another promising avenue of research involves the manipulation of gene networks that regulate inflammation and immune response in muscle tissues. Inflammation is a natural and essential part of the healing process, signaling the body to send repair cells to injured areas. However, excessive or prolonged inflammation can be detrimental, leading to impaired recovery and chronic conditions such as tendinitis and arthritis. By fine-tuning the genetic pathways that control inflammation, scientists aim to optimize the healing process, minimizing harmful inflammation while promoting effective tissue repair.

One of the key genes involved in this process is NF-kB, a protein complex that plays a crucial role in regulating immune response and inflammation. Overactivation of NF-kB can result in excessive inflammation, contributing to tissue damage and delayed healing. Researchers are exploring ways to modulate the activity of NF-kB and other related genes to achieve a balanced inflammatory response. This could involve using CRISPR technology to precisely edit the DNA sequences that control these genes, reducing their activity in cases of excessive inflammation or enhancing their function when a stronger response is needed for effective healing.

Additionally, scientists are investigating the role of cytokines, small proteins that act as signaling molecules in the immune system. Cytokines such as interleukin-6 (IL-6) and tumor necrosis factor-alpha (TNF-α) are known to be involved in the inflammatory response. By manipulating the genes that encode these cytokines, researchers hope to create a more controlled inflammatory environment that supports muscle regeneration and repair. For example, reducing the expression of pro-inflammatory cytokines while increasing the production of anti-inflammatory cytokines could help accelerate the healing process and prevent the development of chronic inflammatory conditions.

Another critical aspect of this research delves into the intricate relationship between inflammation and muscle stem cells, commonly referred to as satellite cells. These satellite cells play a pivotal role in muscle regeneration and repair, activating in response to muscle damage to proliferate and differentiate into mature muscle fibers, thus facilitating recovery and growth. However, chronic inflammation—a persistent and prolonged inflammatory response—can severely impair the functionality of these satellite cells. This impairment manifests as reduced proliferation and differentiation capacity, ultimately hindering effective muscle recovery and regeneration.

Scientists are now focusing on identifying and manipulating the genetic pathways that govern both inflammation and satellite cell activity to counteract these detrimental effects. By gaining a deeper understanding of these pathways, researchers aim to enhance the regenerative capacity of satellite cells. This involves a dual approach: promoting the expression of genes that support satellite cell proliferation and differentiation, while simultaneously suppressing the genes that contribute to chronic inflammation and inhibit satellite cell function.

One promising strategy is to upregulate the expression of anti-inflammatory cytokines and other regulatory proteins that facilitate a conducive environment for satellite cell activity. For instance, genes like IL-10 and TGF-beta, which have anti-inflammatory properties, could be targeted to reduce inflammation and create a more favorable environment for muscle repair. Additionally, researchers are exploring ways to enhance the expression of growth factors such as IGF-1 (Insulin-like Growth Factor 1) and FGF (Fibroblast Growth Factor), which are known to stimulate satellite cell proliferation and differentiation.

On the flip side, efforts are being made to downregulate or inhibit the activity of pro-inflammatory genes and pathways that contribute to chronic inflammation. This could involve targeting

NF-kB, a key regulator of inflammatory responses, or other pro-inflammatory cytokines like TNF-alpha and IL-6, which are often elevated in chronic inflammatory conditions and are known to impair satellite cell function.

In parallel to these advancements, studying animals that can regenerate new limbs, such as salamanders and certain species of lizards, provides valuable insights into the genetic and cellular mechanisms underlying regeneration. These animals possess remarkable abilities to regrow complex tissues, including bones, muscles, nerves, and skin, after injury. By understanding these natural regenerative processes, scientists aim to apply similar principles to human medicine, potentially using CRISPR technology to enable limb regeneration in soldiers who have suffered combat-related amputations.

In regenerative species, injury triggers the activation of a specialized type of stem cells known as blastemal cells. These cells proliferate and differentiate into the various cell types needed to form a new limb. Regeneration also involves complex signaling pathways that guide the growth and differentiation of cells. Key pathways include Wnt, FGF (Fibroblast Growth Factor), and TGF-beta (Transforming Growth Factor-beta), which play crucial roles in tissue development and regeneration. Unlike humans, regenerative animals heal wounds without forming scar tissue, which can impede regrowth. Understanding how these animals avoid scarring can help develop strategies to enhance human tissue regeneration.

By identifying the specific genes involved in limb regeneration in animals, scientists can use CRISPR to edit corresponding genes in humans. For instance, genes that regulate stem cell activation and proliferation could be targeted to promote the formation of blastema-like cells in amputated limbs. CRISPR can be used to enhance or activate the signaling pathways that are critical for regeneration. For example, upregulating genes in the Wnt and FGF pathways could stimulate the growth and differentiation of cells required for limb regeneration. To mimic the scar-free healing seen in regenerative animals, CRISPR could be used to downregulate genes involved in fibrosis and scar tissue formation. This approach aims to create a more conducive environment for tissue regrowth.

Researchers could use CRISPR to reprogram existing cells in the stump of an amputated limb to revert to a more primitive, stem-cell-like state, capable of differentiating into various cell types needed for limb regeneration. However, the complexity of human limbs presents a significant challenge, as human limbs are complex structures with multiple tissue types that need to be regenerated in a coordinated manner. Ensuring that all these tissues grow correctly and integrate seamlessly remains a significant challenge. Additionally, the human immune system might react to the newly formed tissues, potentially leading to rejection or other complications. Strategies to modulate the immune response will be crucial for successful regeneration.

Research is ongoing to explore the feasibility of using CRISPR for limb regeneration. Scientists are conducting studies in animal models to refine gene-editing techniques and identify the most promising targets for enhancing regenerative capabilities. Advances in stem cell biology, tissue engineering, and regenerative medicine are likely to complement these efforts, paving the way for innovative treatments that could one day enable soldiers and civilians to regenerate lost limbs after severe injuries. By leveraging the natural regenerative mechanisms observed in animals and applying cutting-edge CRISPR technology, researchers hope to unlock the potential for human limb regeneration, offering new hope for individuals affected by limb loss.

The potential applications of this research extend beyond injury recovery. Optimizing the inflammatory response could also benefit individuals with chronic inflammatory diseases and those undergoing intense physical training or rehabilitation. For athletes and military personnel, this could mean faster recovery times, reduced risk of chronic injuries, and improved overall performance. In the context of genetic engineering, manipulating inflammation and immune response genes could be a key component of creating enhanced soldiers capable of withstanding the physical demands of modern warfare.

The integration of these advanced genetic modifications could lead to soldiers who not only possess enhanced physical attributes but also exhibit remarkable resilience and endurance. This could fundamentally change the dynamics of military engagements, as enhanced soldiers would be able to withstand greater physical challenges, recover swiftly from injuries, and sustain prolonged operations with minimal downtime. The implications of such advancements are profound, potentially giving militaries a strategic edge by maintaining a highly capable and robust fighting force.

Beyond immediate battlefield applications, the potential for CRISPR to enhance physical capabilities could revolutionize military training programs. Soldiers could undergo genetic modifications early in their training, optimizing their physical development to meet the rigorous demands of military service. This could result in a new generation of soldiers who are not only stronger and more enduring but also more resilient to the physical stresses of combat.

A key area of interest in this context is the enhancement of speed and reflexes, essential attributes in military operations that often determine the outcome of high-stakes engagements. The ability to react swiftly and move with agility can be the difference between life and death on the battlefield. One gene that has garnered significant attention in this context is ACTN3, commonly referred to as the "sprinter gene." This gene is known for its role in enhancing muscle performance, particularly in elite athletes who excel in sprinting and explosive movements.

Research has shown that variations in the ACTN3 gene can influence muscle fiber composition, favoring the development of fast-twitch fibers that are crucial for quick, powerful bursts of activity. By targeting and modifying this gene, scientists could potentially create soldiers with significantly enhanced sprinting abilities and faster reaction times. Such genetic modifications could translate into a formidable tactical advantage, allowing these enhanced soldiers to outmaneuver and outpace their adversaries in various combat scenarios.

Imagine a battlefield where genetically modified soldiers can dash across open terrain with incredible speed, quickly closing the distance between themselves and their targets. These soldiers could execute rapid assaults, perform evasive maneuvers, and reposition themselves with unparalleled efficiency. In close-quarters combat, enhanced reflexes would enable them to respond to threats almost instantaneously, dodging attacks and countering with precise, decisive actions.

Furthermore, the implications extend beyond individual performance. Units composed of soldiers with genetically enhanced speed and reflexes could operate with greater synchronization and effectiveness. Quick, coordinated movements could facilitate complex tactical maneuvers, such as flanking enemy positions or conducting rapid extraction missions under fire.

Enhanced speed would also be advantageous in rescue operations, enabling soldiers to reach and evacuate wounded comrades more swiftly.

The potential benefits of genetic modifications extend far beyond physical confrontations on the battlefield. Enhanced reflexes and faster reaction times could significantly improve performance in various high-pressure situations. For instance, pilots operating advanced aircraft could benefit from increased cognitive and motor skills, allowing them to make split-second decisions with greater accuracy and precision. This could enhance their ability to navigate complex aerial maneuvers, engage in dogfights, and respond rapidly to threats, ultimately increasing their survival rates and mission success.

In addition to piloting aircraft, these enhancements could prove invaluable in operating advanced weapon systems. Soldiers with genetically enhanced reflexes and processing speeds could manage sophisticated technology with ease, ensuring more accurate targeting and more effective use of complex systems such as missile defense arrays, unmanned aerial vehicles (UAVs), and cyber warfare tools. This could provide a strategic advantage by allowing quicker and more efficient responses to enemy actions, reducing the risk of errors and maximizing the effectiveness of military operations. Soldiers with these enhancements would be better equipped to handle the demands of modern warfare, where split-second decisions and rapid responses are often crucial.

Moving beyond physical enhancements, cognitive abilities represent a significant area of interest for military applications of genetic technology. One particular focus is the gene NR2B, which encodes a subunit of the NMDA receptor in the brain, crucial for synaptic plasticity, learning, and memory. Animal studies have demonstrated that overexpressing NR2B can significantly enhance memory and learning capabilities.

If these findings can be translated to humans, soldiers with enhanced cognitive functions could process information more rapidly, make quicker decisions, and retain crucial training and tactical information more effectively. This genetic enhancement could revolutionize the effectiveness of small military units, where quick thinking and adaptability are paramount.

Enhanced cognitive abilities would allow soldiers to better anticipate enemy movements, adjust strategies on the fly, and operate more autonomously in complex and rapidly changing environments. For example, in high-stress combat situations, the ability to process sensory information quickly and accurately could be the difference between life and death. Enhanced memory and learning capabilities would mean that soldiers could more effectively integrate and apply advanced training techniques, including complex tactics and new technology.

Moreover, such cognitive enhancements could facilitate improved coordination and communication within units. Soldiers with superior cognitive functions might be better at understanding and executing complex orders, interpreting battlefield data, and sharing critical information with their comrades in real-time. This would lead to more cohesive and effective unit operations, with each member fully understanding their role and the overall strategy.

Additionally, the implications of cognitive enhancements extend far beyond the battlefield. Soldiers with enhanced cognitive abilities could serve in specialized roles that demand high levels of concentration, advanced problem-solving skills, and strategic planning capabilities. These enhancements could revolutionize areas such as intelligence analysis, cyber warfare, and logistics, where the ability to process and interpret large amounts of information quickly and accurately is crucial.

In intelligence analysis, for instance, enhanced cognitive functions could enable soldiers to sift through vast quantities of data at unprecedented speeds, identifying patterns and connections that might be missed by unenhanced individuals. This capability is vital in an age where information overload can impede effective decision-making. Enhanced analysts could provide timely and accurate intelligence assessments, contributing to more informed and strategic military operations. The ability to rapidly process satellite imagery, intercepted communications, and other intelligence sources could give military leaders a decisive edge in planning and executing missions.

In the realm of cyber warfare, cognitive enhancements could be a game-changer. Cyber operations often require quick thinking and adaptability, as well as the ability to anticipate and counter enemy cyber tactics. Enhanced cognitive functions could allow soldiers to develop more sophisticated algorithms, detect and respond to cyber threats in real-time, and manage complex cyber defense systems with greater efficiency. The capacity to outthink adversaries in the digital domain could be just as critical as physical superiority on the battlefield.

Logistics, a cornerstone of military operations, could also benefit significantly from cognitive enhancements. Managing supply chains, coordinating transportation, and ensuring the timely delivery of resources are complex tasks that require meticulous planning and execution. Enhanced cognitive abilities could improve logistical efficiency by optimizing routes, predicting supply needs based on dynamic battlefield conditions, and streamlining the distribution of resources. This could result in more resilient and adaptable supply chains, reducing the risk of logistical failures that could jeopardize missions.

Moreover, cognitive enhancements could improve decision-making under pressure. Soldiers in command roles often have to make rapid decisions in high-stress environments, where the stakes are incredibly high. Enhanced cognitive functions could improve their ability to assess situations, weigh options, and make sound decisions swiftly. This could enhance overall mission success rates and reduce the likelihood of costly mistakes.

To achieve cognitive enhancements, scientists would need to identify and target specific genes such as BDNF (Brain-Derived Neurotrophic Factor), which plays a crucial role in neuroplasticity, the brain's ability to reorganize itself by forming new neural connections. Enhancing BDNF expression can improve learning, memory, and cognitive flexibility. COMT (Catechol-O-Methyltransferase) affects the breakdown of dopamine in the prefrontal cortex, influencing cognitive functions such as working memory and executive function. Modifying COMT activity could enhance these cognitive aspects. GRIN2B (Glutamate Ionotropic Receptor NMDA Type Subunit 2B) is involved in synaptic plasticity and memory formation, so enhancing its expression could lead to improved learning and memory. NRG1 (Neuregulin 1) is involved in synaptic development and plasticity, and altering its expression can enhance cognitive abilities by promoting better synaptic connections. FOXP2 (Forkhead Box P2) is associated with language and speech processing, and enhancing its expression could potentially improve language acquisition and processing skills.

The process of using CRISPR involves designing a guide RNA (gRNA) that matches the specific DNA sequence of the target gene. The gRNA directs the CRISPR-Cas9 complex to the precise location in the genome. The gRNA is combined with the Cas9 protein, which acts as molecular scissors to cut the DNA at the targeted site. Once the DNA is cut, the cell's natural repair mechanisms kick in. If the goal is to enhance gene function, a DNA template with the desired sequence can be introduced to guide the repair process through homology-directed repair. This template could include sequences that increase the expression or function of the target gene. CRISPR could be used to knock-in sequences that boost gene expression or to knock-out sequences that inhibit gene activity.

To enhance cognitive functions, CRISPR could be used to upregulate BDNF by inserting promoter sequences that increase its expression, thereby enhancing neuroplasticity, learning, and memory. For COMT, CRISPR could create a variant of the gene associated with slower dopamine breakdown, resulting in improved working memory and executive function. Enhancing GRIN2B expression could improve synaptic plasticity and memory formation, while using CRISPR to insert regulatory elements that enhance NRG1 expression could promote better synaptic connections and cognitive performance. Enhancing FOXP2 expression through CRISPR could improve language processing and communication skills.

The potential benefits of cognitive enhancement through genetic modification highlight a new frontier in military science, where the boundaries of human capability are continually pushed. However, the ethical implications of such enhancements must be carefully considered. The prospect of creating soldiers with enhanced cognitive functions raises questions about consent, the potential for abuse, and the long-term effects on individuals and society.

Resilience to stress and fatigue is another crucial factor for soldiers in combat. Modern warfare often demands prolonged periods of alertness and performance under extreme conditions, where the ability to maintain mental acuity and physical endurance can mean the difference between success and failure. Genes associated with circadian rhythms and stress responses, such as CLOCK and NR3C1, have emerged as promising targets for genetic enhancement.

The "clock gene" refers to a group of genes, including CLOCK, BMAL1, PER1, PER2, PER3, CRY1, and CRY2, that play a crucial role in regulating circadian rhythms. These rhythms influence sleep-wake patterns, hormone release, body temperature, and other vital functions. By interacting in a feedback loop, these genes maintain the body's internal clock, synchronizing it with external environmental cues such as light and darkness.

Using CRISPR technology to manipulate clock genes in soldiers could offer significant benefits, particularly for those involved in missions that disrupt normal sleep patterns, such as night operations, long-duration flights, or deployments in extreme environments. Enhancing or modulating the function of clock genes could help soldiers adapt more quickly to changing time zones and shift work, reducing the impact of jet lag and improving cognitive performance and alertness during night missions or extended operations. Additionally, enhancing the function of certain clock genes might increase the efficiency of sleep, allowing soldiers to achieve restorative sleep more quickly, which is crucial in combat situations where rest periods are limited.

Moreover, regulating circadian rhythms more precisely could help optimize the timing of peak physical and cognitive performance, ensuring that soldiers are at their best during critical phases of a mission. Better synchronization of circadian rhythms might also help reduce stress and improve resilience to the psychological and physiological challenges of combat.

For instance, the CLOCK gene encodes a protein that forms a complex with another protein encoded by BMAL1, which then activates the transcription of other clock genes such as PER and CRY. By using CRISPR to enhance the expression of CLOCK or BMAL1, scientists could potentially stabilize circadian rhythms, improving sleep quality and duration.

However, there are significant risks and ethical considerations associated with manipulating clock genes. Disrupting natural circadian rhythms can have unintended consequences, including metabolic disorders, mood disturbances, and impaired immune function. The long-term effects of such genetic modifications are not fully understood, and there is a risk of off-target effects where unintended genes are edited, leading to unpredictable outcomes.

Ethically, manipulating clock genes raises concerns about consent, autonomy, and the potential for coercion. Soldiers might feel pressured to undergo genetic modifications to meet military demands, leading to ethical dilemmas about personal freedom and the right to refuse such interventions.

Similarly, the NR3C1 gene, which codes for the glucocorticoid receptor, is integral to the body's response to stress. Glucocorticoids, including cortisol, are hormones that help regulate metabolism, immune response, and stress reactions. These hormones are crucial for mobilizing energy, suppressing inflammation, and maintaining homeostasis during stressful situations. When the body perceives stress, the hypothalamus releases corticotropin-releasing hormone (CRH), which in turn triggers the release of adrenocorticotropic hormone (ACTH) from the pituitary gland. ACTH then stimulates the adrenal cortex to release glucocorticoids.

Enhancing the function of the NR3C1 gene could potentially amplify the sensitivity and efficiency of the glucocorticoid receptors, leading to a more effective stress response. This genetic modification might result in soldiers having an enhanced ability to remain calm and focused in high-pressure situations. With more efficient glucocorticoid signaling, soldiers could better manage the physiological impacts of stress, maintaining cognitive function and decision-making abilities even under extreme duress.

Furthermore, enhanced glucocorticoid receptor function could lead to faster recovery times from stressful encounters. This means that soldiers would not only perform better during high-stress situations but also recover more quickly afterward, reducing the cumulative toll that prolonged stress can take on the body. This could be particularly beneficial in combat scenarios where soldiers are exposed to repeated and intense stressors.

Reducing the likelihood of stress-related disorders, such as anxiety and depression, is another significant potential benefit. Chronic stress is a known risk factor for these conditions, and enhancing the body's ability to manage and recover from stress could mitigate this risk. Studies have shown that individuals with more robust glucocorticoid signaling pathways are less susceptible to stress-induced mood disorders. By enhancing NR3C1 gene function, it might be possible to bolster the mental health resilience of soldiers, providing a safeguard against the psychological impacts of combat and prolonged stress.

In addition to these mental health benefits, enhanced stress resilience could improve overall physical health. Chronic stress has been linked to numerous health issues, including cardiovascular disease, weakened immune function, and metabolic disorders. By improving the body's stress response, soldiers could maintain better overall health, reducing downtime due to illness and improving long-term well-being.

Imagine this hypothetical story about a military sniper named Jake. One night, Jake and his team were deployed on a critical mission deep in enemy territory. The operation required them to remain awake, alert and undetected for over 48 hours, waiting for the perfect moment to strike. As the hours dragged on, Jake's enhanced clock genes kept him alert and focused. His team relied on his precise instructions and unwavering calm, both products of his enhanced NR3C1 gene.

When the time finally came to take the shot, Jake's enhanced vision and reflexes—thanks to the optimized circadian rhythms—ensured that he hit his target with pinpoint accuracy. The mission was a success, and the team exfiltrated without incident.

By targeting and modifying these genes, the military could also potentially develop soldiers who are not only more resistant to the physical and psychological strains of combat but also require less rest and recovery time. This genetic enhancement could produce a force that is perpetually ready and capable of sustained high performance under the most demanding conditions. Imagine a scenario where soldiers can operate for days on end without significant drops in performance, maintaining sharp decision-making skills and physical prowess even in the harshest environments.

Furthermore, the implications of such genetic modifications extend beyond immediate combat readiness. Enhanced stress resilience and altered circadian rhythms could also improve overall health and longevity, reducing the long-term impacts of service on soldiers' bodies and minds. This would not only benefit military operations but also the well-being of service members during and after their military careers.

As we explore these possibilities, it becomes evident that the line between human and machine begins to blur. The integration of genetic enhancements with cybernetic technologies could create soldiers who are not only genetically optimized but also augmented with advanced technology. This convergence of biology and technology could redefine what it means to be a soldier, raising further questions about the nature of humanity and the ethics of creating enhanced individuals for the purpose of warfare.

As we conclude this section, it's clear that while the potential genetic enhancements for soldiers present remarkable opportunities, they also bring with them significant ethical and practical challenges. These enhancements could revolutionize military capabilities, but they must be pursued with a deep understanding of the consequences and a commitment to ethical principles. Moving forward, we will explore the broader ethical and moral dilemmas posed by these advancements, and how they intersect with our understanding of humanity and societal values.

Chapter 4.

Ethical and Moral Dilemmas

As we delve deeper into the realm of genetic modifications in humans, one cannot ignore the profound ethical questions that arise. The promise of CRISPR technology is undoubtedly transformative, offering the potential to eradicate genetic diseases, enhance human abilities, and perhaps even extend life. However, these possibilities come with significant moral and ethical considerations that society must grapple with.

One of the foremost ethical dilemmas is the concept of "playing God." This notion revolves around the fundamental question of whether humans should have the power to alter their genetic makeup at will. Historically, humanity has sought to improve itself through various means—medicine, education, and technology. However, genetic modification represents a paradigm shift, allowing for changes at the most basic level of human existence.

The potential to eliminate hereditary diseases is undeniably positive. For instance, CRISPR has shown immense promise in treating a variety of genetic conditions, including sickle cell anemia and cystic fibrosis, offering hope for cures to previously untreatable diseases. In 2019, scientists made headlines by using CRISPR to treat a patient with sickle cell disease, marking a significant milestone in medical history. This groundbreaking procedure involved extracting hematopoietic stem cells from the patient, editing them outside the body using CRISPR to correct the defective gene responsible for sickle cell disease, and then reintroducing these modified cells back into the patient's bloodstream. The edited cells were able to produce healthy red blood cells, significantly alleviating the symptoms of the disease.

The success of this treatment not only provided a new lease on life for the patient but also demonstrated the transformative potential of CRISPR-based therapies. This approach offers a more precise and permanent solution compared to traditional treatments, which primarily focus on managing symptoms rather than addressing the root cause of genetic disorders.

Similarly, CRISPR has been used to target the genetic mutations that cause cystic fibrosis, a debilitating condition affecting the lungs and digestive system. By correcting the defective CFTR gene in the affected cells, researchers aim to restore normal function and prevent the severe complications associated with the disease. Early experiments in laboratory settings and animal models have shown promising results, paving the way for future clinical trials and potential treatments for patients suffering from cystic fibrosis.

These advancements in CRISPR technology represent a paradigm shift in the field of genetic medicine, offering new possibilities for curing genetic diseases at their source. As researchers continue to refine these techniques and expand their applications, the hope is that CRISPR will become a standard tool in the fight against a wide array of genetic disorders, ultimately transforming the landscape of modern medicine and improving the lives of millions worldwide. Yet, this capability also raises the specter of eugenics—the controversial idea of selectively breeding humans to enhance desirable traits and eliminate undesirable ones.

The dark history of eugenics, particularly during the early 20th century, serves as a sobering cautionary tale of scientific hubris and moral failure. Eugenics programs in countries like the United States and Germany led to horrific practices, including forced sterilizations and the systematic persecution of individuals deemed "unfit." These actions were justified by the deeply flawed and dangerous belief that society could be improved by eradicating certain traits and populations. In the United States, the eugenics movement led to the forced sterilization of tens of thousands of people, often targeting those with mental illnesses, disabilities, and other marginalized groups. The Supreme Court case Buck v. Bell in 1927 notoriously upheld the constitutionality of these sterilizations, with Justice Oliver Wendell Holmes Jr. declaring, "Three generations of imbeciles are enough."

In Germany, the eugenics ideology underpinned the Nazi regime's atrocities, including the genocide of millions during the Holocaust. The Nazis implemented their own brutal eugenics program, which involved the forced sterilization and euthanasia of those they deemed "racially inferior" or "genetically defective." These inhumane policies were justified under the guise of purifying the Aryan race and eliminating perceived genetic threats to societal health.

The advent of CRISPR technology, with its unprecedented precision and accessibility, brings with it the potential for both remarkable medical breakthroughs and significant ethical dilemmas. The ability to edit genes with such accuracy could inadvertently revive these dangerous ideologies if not carefully regulated and ethically guided. CRISPR could be misused to promote modern forms of eugenics, targeting traits deemed undesirable and further entrenching social inequalities. The allure of creating so-called "designer babies," free from perceived imperfections, echoes the discredited beliefs of the early eugenicists who sought to engineer a "better" humanity through selective breeding and genetic intervention.

The ethical question, therefore, is whether the potential benefits of genetic modification justify the risks of reviving such harmful practices. As we stand on the brink of a new era in genetic engineering, it is crucial to remember the lessons of the past. We must ensure that the use of CRISPR and similar technologies is guided by principles of equity, justice, and respect for human dignity. This includes establishing robust regulatory frameworks to prevent abuse, fostering public discourse on the ethical implications, and promoting transparency in genetic research and its applications.

Only by confronting the ethical challenges head-on and committing to responsible stewardship can we harness the power of CRISPR for the benefit of all humanity, without repeating the grave mistakes of the past.

Moreover, the notion of genetic enhancement introduces another layer of ethical complexity. Beyond curing diseases, CRISPR technology holds the potential to enhance human abilities—ranging from physical attributes like strength and intelligence to less tangible traits such as personality and behavior. This possibility raises questions about equity and fairness. If genetic enhancements become available, who gets access to them? Will they be a luxury only available to the wealthy, thus exacerbating existing social inequalities? The ethical implications of creating a genetically enhanced elite class are profound and troubling. It could lead to a society where natural genetic diversity is devalued, and people are judged based on their genetic modifications rather than their inherent human qualities.

The potential for unintended consequences also cannot be overlooked. Genetic modifications can have unforeseen effects, leading to new health issues or exacerbating existing ones. For instance, a modification intended to enhance cognitive abilities could inadvertently increase the risk of mental health disorders. The long-term effects of genetic modifications on the human gene pool are still largely unknown, and the ethical responsibility to future generations must be considered. This uncertainty underscores the need for rigorous oversight and long-term studies before widespread adoption of genetic modification technologies.

In addition, the issue of consent is particularly thorny when it comes to genetic modifications. Adults may be able to give informed consent for themselves, but what about modifications made to embryos or children? These individuals cannot consent to changes that will affect their entire lives, raising significant ethical questions about autonomy and the rights of the individual. The case of He Jiankui, a Chinese scientist who announced in 2018 that he had created the world's first genetically edited babies, sparked international outrage and condemnation. He claimed to have used CRISPR technology to modify the embryos of twin girls to confer resistance to HIV, a groundbreaking but highly controversial endeavor. The announcement was met with swift and severe backlash from the global scientific community, bioethicists, and the public, highlighting a myriad of ethical breaches and concerns.

Firstly, one of the most significant ethical violations was the lack of proper informed consent from the parents involved in the experiment. Reports indicated that the parents were not fully informed of the potential risks and implications of the genetic modifications, raising serious questions about their understanding and voluntary participation. This breach of ethical standards undermines the trust required for any medical or scientific intervention, especially one as profound and irreversible as genetic editing.

Furthermore, the potential long-term risks to the genetically modified children remain largely unknown. CRISPR technology, while precise, is not infallible and can result in off-target effects—unintended changes to other parts of the genome that could lead to unforeseen health issues, including cancer or other genetic disorders. The modifications introduced into these embryos will not only affect the twins but could also be passed down to future generations, compounding the ethical dilemma with the prospect of heritable genetic changes.

The case also underscored the absence of robust ethical and regulatory frameworks to govern such groundbreaking research. While the potential benefits of genetic editing in preventing diseases are immense, the lack of comprehensive guidelines and oversight makes the rush to apply these technologies particularly dangerous. He Jiankui's actions bypassed established scientific protocols and ethical norms, prompting urgent calls for the development of international standards to ensure responsible research and application of genetic technologies.

In response to the outcry, Chinese authorities sentenced He Jiankui to three years in prison for illegal medical practices, and the scientific community called for a moratorium on clinical uses of human germline editing. This case has ignited a global debate on the ethical boundaries of genetic engineering, the need for stringent oversight, and the responsibilities of scientists to prioritize the welfare and rights of individuals over the pursuit of scientific advancement.

The He Jiankui case serves as a stark reminder of the perils of unchecked scientific experimentation and the critical need for ethical vigilance. It highlights the importance of ensuring that scientific progress does not outpace the development of ethical frameworks and regulatory measures designed to protect human subjects and preserve public trust in biomedical research. As the capabilities of CRISPR and other gene-editing technologies continue to expand, the lessons from this case must inform future research to navigate the fine line between innovation and ethical responsibility.

The cultural and societal implications of genetic modification are profound and complex, demanding careful consideration and respect for diverse perspectives. Different cultures have varying viewpoints on the acceptability and ethics of altering human genetics, influenced by deep-seated beliefs, religious doctrines, and historical contexts. For instance, in some cultures, genetic modification might be viewed as an affront to natural or divine laws, a dangerous overstep that challenges the fundamental principles of life as governed by nature or a higher power. This perspective often stems from religious teachings that emphasize the sanctity and inviolability of the human body, as well as philosophical views that stress the importance of maintaining the natural order.

Conversely, other cultures may welcome genetic modification as a groundbreaking advancement that holds the promise of alleviating human suffering and enhancing quality of life. In these societies, the potential to eradicate genetic diseases, extend lifespans, and improve human capabilities is seen as a logical progression of scientific and technological development. This embrace of genetic engineering is often rooted in a forward-looking mindset that values innovation and the continual improvement of human conditions through science and technology.

These cultural differences must be respected and considered in the global discourse on genetic modification. As genetic technologies become increasingly accessible and their applications more widespread, it is essential to foster international cooperation and dialogue. This collaborative approach can help ensure that the development and implementation of genetic modification technologies are guided by ethical guidelines that honor cultural diversity and protect human rights.

Creating a global framework for the ethical use of genetic modification involves addressing several key issues. First, it requires an understanding of the specific cultural and ethical concerns associated with genetic engineering in different regions. Engaging with local communities, religious leaders, ethicists, and policymakers is crucial to capturing the full spectrum of opinions and values.

Second, it necessitates the establishment of international standards that balance respect for cultural differences with the need to protect individuals from harm and exploitation. These standards should include stringent regulations to prevent misuse of genetic technologies, safeguards to ensure informed consent, and mechanisms to address potential social and economic inequalities that may arise from genetic enhancements.

Third, ongoing education and public engagement are vital to foster an informed and inclusive debate about genetic modification. By raising awareness and facilitating discussions, stakeholders can better understand the benefits, risks, and ethical implications of genetic technologies, enabling societies to make informed decisions that reflect their values and priorities.

Ultimately, the goal is to develop a set of ethical guidelines that are universally respected yet flexible enough to accommodate cultural diversity. Such guidelines should emphasize the importance of human dignity, autonomy, and justice, ensuring that the advancements in genetic modification contribute to the well-being of all humanity rather than exacerbating existing inequalities or creating new forms of discrimination. By prioritizing international cooperation and respectful dialogue, we can navigate the ethical landscape of genetic modification in a way that honors our shared humanity and diverse cultural heritage.

As we consider these ethical questions, it is clear that the path forward requires careful deliberation and robust ethical frameworks. The potential benefits of CRISPR technology are immense, but they must be weighed against the moral and ethical risks. Society must engage in an ongoing dialogue to navigate these complex issues, ensuring that the application of genetic modifications in humans is guided by principles of equity, justice, and respect for human dignity.

The ethical landscape of genetic modifications is intricate and fraught with challenges, but it is not insurmountable. By addressing these questions with seriousness and rigor, we can hope to harness the power of CRISPR technology in a way that benefits humanity while safeguarding against its potential dangers. As we move forward, it is imperative that we remain vigilant and thoughtful, ensuring that our pursuit of scientific advancement does not come at the cost of our ethical integrity.

The discussion of ethical dilemmas is far from over, and it seamlessly leads us to the next critical aspect of this debate: the geopolitical ramifications of CRISPR technology. How will nations navigate the delicate balance between scientific progress and international stability? What regulations and agreements are necessary to prevent a global arms race fueled by genetic advancements? The answers to these questions will shape the future of not only warfare but also the global order as we know it.

Bioethicists are particularly vocal about the moral consequences of genetic modification. They argue that altering human DNA, especially for military purposes, crosses a line that could lead to unforeseen and potentially dangerous outcomes. The central ethical dilemma revolves around the concept of "playing God" and the impact on human identity and agency. Bioethicist George Annas has pointed out that genetic modifications could erode what it means to be human, creating individuals who may possess enhanced physical or cognitive abilities but at the cost of their humanity and individuality.

This concern is not merely theoretical. The history of eugenics serves as a sobering reminder of the potential for misuse. In the early 20th century, eugenics movements in various countries, including the United States and Nazi Germany, aimed to improve the human race through selective breeding. These efforts led to human rights abuses, forced sterilizations, and, in the case of Nazi Germany, the horrors of the Holocaust. Modern bioethicists fear that CRISPR could be a tool for a new form of eugenics, one that is technologically advanced but equally dangerous. Ruth Faden, a prominent bioethicist, argues that while CRISPR offers the possibility of eliminating genetic diseases, it also opens the door to genetic discrimination and new forms of inequality.

Scientists, on the other hand, delve deeply into the technical aspects and potential benefits of CRISPR technology, balancing their enthusiasm with a recognition of the accompanying ethical concerns. They often highlight the transformative promise of genetic editing to eradicate debilitating diseases and significantly enhance human capabilities. For instance, Jennifer Doudna, one of the pioneers of CRISPR technology, has extensively discussed its potential to cure genetic disorders like cystic fibrosis and sickle cell anemia. Her work underscores the profound impact CRISPR could have on improving human health and reducing suffering caused by genetic diseases.

Despite these optimistic views, Doudna and her colleagues are acutely aware of the ethical pitfalls. They argue for the establishment of stringent ethical guidelines to prevent the misuse of such powerful technology. The concern is not just theoretical; the ability to edit human genes raises the specter of eugenics, genetic discrimination, and unintended ecological consequences. The scientific community is thus divided on how to proceed. Some researchers advocate for a moratorium on human genetic editing until we gain a clearer understanding of the long-term effects. They emphasize the need for extensive research on potential off-target effects, unintended consequences, and the ethical implications of germline editing, which affects future generations.

Others in the scientific community push for the continuation of research under strict regulatory frameworks. They argue that halting progress could delay critical medical advancements and leave numerous diseases untreatable. These proponents believe that with proper oversight, the benefits of CRISPR can be harnessed safely and effectively. They propose robust regulatory measures, including international collaboration to ensure ethical standards are met globally.

This division within the scientific community highlights the complexity of balancing innovation with ethical responsibility. As CRISPR technology continues to advance, it will be crucial for scientists, ethicists, policymakers, and the public to engage in ongoing dialogue. This will help navigate the fine line between groundbreaking medical progress and the potential for ethical transgressions, ensuring that CRISPR's benefits are maximized while minimizing its risks.

Military officials offer yet another perspective, driven by the strategic advantages that genetically modified soldiers could provide. The potential to create soldiers with enhanced strength, endurance, and cognitive abilities is a tantalizing prospect for any military force. Enhanced soldiers could operate for longer periods without fatigue, carry heavier equipment, and process information at speeds far surpassing their unmodified counterparts. Such capabilities would not only enhance individual soldier performance but also significantly increase the overall effectiveness of military operations.

General Robert Brown, a retired U.S. Army officer, has spoken about the need to maintain a technological edge over potential adversaries. In numerous interviews and public statements, he has emphasized that staying ahead in technological advancements is crucial for national security. Brown points out that adversaries are not standing still; they are also investing heavily in emerging technologies, including genetic engineering. In this context, the prospect of genetic enhancements for soldiers becomes not just an opportunity but a necessity to maintain parity or gain superiority.

General Brown envisions a future where genetic modifications could provide soldiers with superior physical attributes, such as increased muscle mass and bone density, which would enhance their ability to carry out physically demanding tasks. Additionally, improvements in cognitive functions could lead to quicker decision-making, better problem-solving skills, and enhanced situational awareness on the battlefield. These traits could be crucial in high-stakes scenarios where split-second decisions can determine the outcome of engagements.

However, Brown also acknowledges the ethical minefield this technology presents. The idea of genetically modifying humans, especially for combat purposes, raises profound ethical questions.

Issues such as consent, the potential for unintended consequences, and the long-term effects on individuals and society must be carefully considered. Brown stresses the importance of developing comprehensive policies to govern the use of genetic enhancements in the military. He advocates for an international dialogue to establish guidelines and regulations that ensure these technologies are used responsibly and ethically.

Brown's perspective is not isolated. Many military strategists and ethicists are calling for a balanced approach that weighs the strategic advantages against the ethical implications. They emphasize the need for transparent research practices, informed consent from those undergoing genetic modifications, and robust oversight mechanisms to prevent abuse and ensure accountability.

The potential strategic advantages of genetically modified soldiers are clear, but they come with significant ethical and regulatory challenges. As General Brown and other military officials suggest, the path forward requires careful consideration, international cooperation, and the development of policies that balance innovation with ethical responsibility. By addressing these challenges proactively, the military can harness the benefits of genetic enhancements while minimizing the risks and ensuring that technological advancements serve the greater good.

The convergence of these viewpoints underscores the complexity of using CRISPR technology in the military. Each perspective brings to light different facets of the ethical, technical, and strategic challenges. The debate is not just about the capabilities of the technology but about the values and principles that should guide its application.

In a perfect world, the implementation of CRISPR genetic modifications in the military would be entirely voluntary, respecting the autonomy and personal choice of each soldier. Soldiers would be thoroughly educated on the potential benefits and risks associated with genetic enhancements, ensuring they could make informed decisions without any coercion or undue influence. Comprehensive consent processes would be in place, guaranteeing that every participant fully understands the long-term implications of the modifications. The military would prioritize the well-being and rights of its personnel, offering robust support systems, including medical monitoring and psychological counseling, to assist those who choose to undergo the modifications. This approach would foster a culture of trust and respect, where technological advancements are balanced with ethical considerations, and the enhancement of human capabilities is pursued in a manner that upholds the dignity and freedom of every individual.

Implementing CRISPR gene modification in the military would be a complex process, involving several critical steps and considerations. Initially, extensive research and development would be necessary. This would involve preclinical studies in animal models, followed by clinical trials to test the safety and efficacy of the gene modifications in humans. Collaboration with academic institutions, biotech companies, and regulatory agencies would be crucial during this phase.

Before any implementation, a robust ethical and legal framework would need to be established. This would include guidelines for consent, privacy, and the handling of genetic information. International agreements might also be necessary to ensure compliance with global standards. Unlike mandatory vaccinations, implementing CRISPR gene modifications would likely require explicit informed consent from the soldiers. They would need to be fully briefed on the potential benefits, risks, and long-term implications of the procedure. This process would ensure that soldiers voluntarily agree to the modifications, understanding all the potential consequences.

Comprehensive health and safety protocols would need to be developed to monitor and manage any adverse effects. This would include regular medical check-ups, long-term health monitoring, and psychological support to address any physical or mental health issues that may arise from the gene modifications. The military might initially implement CRISPR modifications through pilot programs, selecting a small, voluntary group of soldiers to undergo the procedure. The results from these pilot programs would be closely monitored and analyzed to refine the process and address any unforeseen challenges.

Extensive training and education programs would be necessary to prepare medical personnel and soldiers for the gene modification process. This would ensure that everyone involved understands the procedures, potential risks, and benefits. Gaining regulatory approval from bodies like the FDA (Food and Drug Administration) in the United States would be essential. This approval would validate the safety and efficacy of the CRISPR modifications, providing a legal basis for their implementation in the military.

Given the ethical and health implications, it is likely that CRISPR gene modifications would be offered on a voluntary basis rather than being a mandatory requirement. Soldiers would be given the option to undergo genetic modifications, similar to how experimental treatments or elective medical procedures are handled. Soldiers would need to be fully informed about the potential risks and benefits. This would involve detailed briefings and consultations with medical professionals to ensure that soldiers make well-informed decisions. They would need to understand the short-term risks, such as potential side effects from the procedure, and long-term implications, such as possible genetic changes being passed to offspring.

The military would need to carefully navigate the ethical considerations of implementing genetic modifications. This includes respecting individual autonomy, ensuring that participation is genuinely voluntary, and providing robust support systems for those who choose to participate. Enhanced physical abilities, increased cognitive functions, and improved resilience to environmental stressors could provide significant strategic advantages. These modifications could lead to a more effective and versatile fighting force, capable of performing better in a wide range of conditions.

However, there are significant risks associated with CRISPR gene modifications, including unintended genetic mutations, unknown long-term health effects, and ethical concerns about human enhancement. The psychological impact on soldiers, who might feel pressured to undergo modifications to keep up with their peers, also needs to be considered. Implementing CRISPR gene modification in the military would be a pioneering yet contentious endeavor. It would require a careful balance between advancing military capabilities and upholding ethical standards. The process would need to be transparent, voluntary, and guided by comprehensive research and ethical guidelines to ensure the well-being and autonomy of the soldiers involved.

In a more realistic scenario, CRISPR genetic modifications would not be presented as a voluntary option but rather imposed as a mandatory requirement for soldiers. This approach would be driven by a combination of strategic imperatives and bureaucratic efficiency, reflecting the military's need to maintain technological and operational superiority over potential adversaries. Soldiers would be compelled to undergo genetic modifications as part of their enlistment or deployment processes, framed as a necessary measure to enhance their combat effectiveness and ensure national security. The military would likely justify this mandate by emphasizing the existential threats posed by genetically enhanced enemy forces and the critical need to level the playing field. The process would involve minimal input from the soldiers themselves, with limited opportunities for informed consent or dissent. Any resistance would be met with disciplinary action, effectively coercing compliance. This forced implementation would prioritize collective military strength and readiness over individual autonomy and ethical considerations, presenting the genetic enhancement of soldiers as an unavoidable aspect of modern warfare.

If CRISPR gene modification were to be implemented as a mandatory requirement in the military, the process would take on a more ominous and authoritarian character, marked by propaganda, coercion, and the shadow of global tensions.

The decision to enforce mandatory gene modification would originate from the highest echelons of military and governmental leadership, driven by a relentless desire to maintain a strategic edge over hostile nations. These nations, perceived as threats to national security, might already be mandating genetic enhancements for their soldiers, compelling the US to adopt similar measures to avoid falling behind. This sense of urgency would be conveyed through an aggressive propaganda campaign aimed at convincing both the military personnel and the public of the necessity and righteousness of the program.

Propaganda efforts would emphasize the existential threat posed by genetically enhanced enemy forces, portraying mandatory genetic modification as a patriotic duty and a critical component of national defense. Soldiers would be inundated with messages highlighting the supposed invincibility of enhanced enemies and the dire consequences of lagging behind in the genetic arms race. This relentless messaging would be designed to suppress dissent and foster acceptance, making the mandatory program appear as an inevitable and essential step.

The implementation process would begin with comprehensive medical screenings, ostensibly to ensure the suitability of soldiers for genetic modification. However, these screenings could also serve as a means of identifying and isolating potential dissenters. Soldiers deemed unfit or resistant to the program might face severe repercussions, including reassignment, demotion, or discharge.

Once screened, soldiers would undergo the CRISPR modifications systematically. New recruits, in particular, would be targeted, with the modifications administered during their initial training. This approach would indoctrinate them into the new military paradigm from the outset, minimizing resistance. The genetic modifications would focus on enhancing physical strength, cognitive functions, and resilience to environmental stressors, creating a new breed of super-soldiers.

Medical oversight would be extensive but not necessarily benevolent. Regular check-ups and advanced diagnostics would be employed not just to monitor health, but also to ensure compliance and detect any signs of resistance or side effects that could jeopardize the program's success. Soldiers exhibiting adverse effects might be silenced or used as cautionary tales to deter others from questioning the modifications.

Ethical considerations would be sidelined in favor of a relentless pursuit of military superiority. The military might employ coercive tactics, such as withholding benefits or threatening dishonorable discharge, to ensure compliance. Any semblance of ethical oversight would be perfunctory, designed to placate critics rather than genuinely protect soldiers' rights. External oversight bodies would likely be complicit, rubber-stamping procedures without rigorous scrutiny.

Legal frameworks would be manipulated to support the mandatory program. Emergency powers and national security provisions could be invoked to override existing laws and suppress legal challenges. Soldiers seeking to resist or expose the program might face severe legal repercussions, including imprisonment or court-martial.

Psychological support, if provided, would likely be superficial, aimed more at maintaining operational efficiency than addressing genuine mental health concerns. Counseling services might be used to reinforce propaganda, convincing soldiers of the necessity and benefits of the modifications while minimizing the perceived risks.

Making CRISPR gene modification mandatory in the military would involve a dark and coercive approach, characterized by propaganda, suppression of dissent, and an overriding focus on maintaining strategic superiority. The process would be marked by ethical compromises, legal manipulation, and an oppressive atmosphere that prioritizes national security over individual rights and well-being.

As we move deeper into the 21st century, the discussion around CRISPR and genetic modification in humans will likely intensify. The stakes are high, and the decisions made today will have far-reaching consequences. Balancing the potential benefits with the ethical and moral concerns is crucial. It requires a nuanced approach that considers the insights and warnings from bioethicists, the technical expertise of scientists, and the strategic considerations of military officials. Only through such a multidisciplinary dialogue can we hope to navigate the challenges posed by this powerful technology responsibly.

In the following sections, we will delve into the geopolitical ramifications of CRISPR technology, exploring how the potential for genetically modified soldiers could alter global power dynamics and international relations.

Chapter 5.

The Geopolitical Landscape

The ethical implications of deploying CRISPR-modified soldiers raise significant questions about the future of warfare. Beyond the immediate concerns about the morality of genetic manipulation, the potential for CRISPR soldiers to alter global power dynamics warrants serious consideration. As nations race to harness the power of genetic engineering for military purposes, the geopolitical landscape stands on the brink of profound transformation.

Imagine a scenario where nations possess the capability to enhance their soldiers' physical and cognitive abilities through genetic modifications. These enhancements could range from increased strength and endurance to heightened resistance to diseases and faster recovery from injuries. Such advancements would provide a strategic advantage on the battlefield, allowing countries with CRISPR technology to project power more effectively and sustain prolonged military engagements with fewer casualties.

Historically, technological advancements in warfare have played pivotal roles in shifting the balance of power among nations, often with profound and far-reaching consequences. One of the most significant examples of this dynamic occurred during World War II with the development and deployment of nuclear weapons. The introduction of atomic bombs by the United States not only brought a swift and decisive end to the conflict with Japan but also fundamentally altered the geopolitical landscape. The sheer destructive power of nuclear weapons established the United States as a dominant military force, catapulting it to superpower status.

The aftermath of World War II saw the emergence of the Soviet Union as the other global superpower, largely due to its rapid development of its own nuclear arsenal. This nuclear parity between the United States and the Soviet Union set the stage for the Cold War, a period characterized by intense rivalry and the ever-present threat of nuclear annihilation. The arms race that ensued was driven by the constant pursuit of technological superiority, with both nations investing heavily in the development of increasingly sophisticated and powerful weapons systems.

This historical precedent highlights how technological innovation in warfare can redefine international relations and power structures. The invention of nuclear weapons created a new era of deterrence and mutually assured destruction (MAD), wherein the potential for catastrophic consequences kept direct military conflicts between superpowers at bay. Instead, the Cold War was marked by proxy wars, espionage, and a continuous quest for technological edge, including advancements in missile technology, space exploration, and cyber warfare.

In more recent times, the advent of drone technology and cyber capabilities has continued to reshape military strategy and global power dynamics. Drones have revolutionized modern warfare by providing new means of surveillance and targeted strikes, often with reduced risk to human soldiers. Cyber warfare, on the other hand, has introduced a new domain of conflict where nations can engage in sabotage, espionage, and disruption without physical confrontation.

Looking ahead, the integration of artificial intelligence, robotics, and genetic engineering into military applications promises to further alter the landscape of global power. As nations invest in these cutting-edge technologies, we may witness the emergence of new forms of warfare that blend human and machine capabilities. For instance, the potential creation of genetically enhanced soldiers through technologies like CRISPR could provide unprecedented advantages on the battlefield, raising ethical and strategic questions about the future of combat and the balance of power.

The pattern is clear: each major leap in military technology not only transforms the methods of warfare but also redefines the hierarchy of global power. The development of nuclear weapons during World War II is a stark reminder of how a single technological breakthrough can shape the course of history, ushering in new eras of political tension and competition. As we stand on the brink of another technological revolution in warfare, understanding this historical context is crucial for anticipating the potential implications and preparing for the challenges ahead.

The introduction of CRISPR soldiers could create a new arms race, with countries vying to develop the most advanced genetically modified troops. This competition could lead to significant investments in genetic research and military biotechnology, potentially at the expense of other critical areas such as healthcare, education, and infrastructure.

Moreover, the existence of CRISPR soldiers could complicate international relations and exacerbate existing tensions. Countries that lag in genetic enhancement technology might feel threatened and seek to form alliances or engage in espionage to close the gap. On the other hand, nations with advanced CRISPR capabilities might leverage their technological superiority to coerce or intimidate others, leading to a destabilization of regional and global power structures.

The proliferation of CRISPR technology also raises concerns about its accessibility and potential misuse. Unlike nuclear weapons, which require substantial resources and infrastructure, CRISPR technology is relatively inexpensive and accessible. This democratization of genetic engineering means that not only state actors but also non-state actors, such as terrorist organizations and rogue states, could potentially develop CRISPR-modified soldiers. The implications of such entities possessing enhanced soldiers are alarming, as it could lead to asymmetric warfare scenarios where traditional military forces are ill-equipped to respond.

The advent of CRISPR technology presents both opportunities and significant risks, particularly when considering its potential use in creating biological weapons for war or terrorism. The ability to precisely edit genes could enable the development of weapons that are not only more effective but also more targeted and difficult to detect. The ability to precisely edit genes could enable the development of weapons that are not only more effective but also more targeted and difficult to detect. These genetically engineered bioweapons would be difficult to detect for several reasons.

Genetically engineered pathogens can be designed to closely resemble naturally occurring strains, making it challenging to distinguish between an intentional release and a natural outbreak. This camouflage can delay the identification of the bioweapon and hinder the appropriate response. By using CRISPR technology, bioweapons can be engineered to target specific genetic markers within a population. This specificity means that the pathogen might only affect certain ethnic groups or individuals with particular genetic traits, leading to localized outbreaks that could be initially perceived as isolated health issues rather than an act of bioterrorism.

Some genetically modified pathogens can be designed to have a delayed onset of symptoms, making it harder to trace the source of the infection. If individuals show symptoms weeks or months after exposure, identifying the point of origin becomes significantly more complicated. CRISPR can also be used to engineer pathogens that evade existing diagnostic tools. By altering genetic sequences that diagnostic tests rely on, these pathogens can go undetected in routine screenings, making early detection and containment more challenging.

Engineered pathogens can be designed to survive in a variety of environmental conditions, including extreme temperatures or humidity. This resilience allows them to persist in the environment for longer periods, increasing the window of exposure and complicating efforts to trace and neutralize the bioweapon. CRISPR technology can also be used to create pathogens with atypical symptoms that do not match known diseases, leading to misdiagnosis and inappropriate treatments. This misdirection can delay the identification of the true nature of the outbreak and the implementation of effective countermeasures.

Genetically modified pathogens can be designed to mutate rapidly, adapting to host defenses and treatments. This adaptability makes it harder to develop vaccines and treatments, prolonging the time it takes to identify and counteract the bioweapon. The combination of these factors makes genetically engineered bioweapons particularly insidious. The ability to precisely edit genes allows for the creation of pathogens that blend seamlessly with natural disease processes, evade detection, and target specific populations, making them formidable tools for covert biological warfare and terrorism. This underscores the need for enhanced surveillance, rapid diagnostic tools, and international cooperation to detect and respond to such threats effectively.

Building on this potential for harm, CRISPR can be used to modify the genomes of bacteria and viruses to enhance their virulence, transmissibility, and resistance to existing treatments. For example, a terrorist group could engineer a strain of the influenza virus to be more deadly and resistant to antiviral drugs. This modified pathogen could spread rapidly through a population, causing widespread illness and death before effective countermeasures could be deployed.

Using CRISPR, it is possible to design bioweapons that target specific genetic profiles, raising chilling possibilities for the future of warfare and terrorism. This technology could enable the creation of pathogens engineered to affect individuals with particular genetic markers while leaving others unharmed. The implications of such bioweapons are profound, as they could be used to target specific ethnic groups or populations, leading to ethnic cleansing or selective genocide. The potential for abuse is staggering, given that these bioweapons could be deployed covertly, masquerading as natural outbreaks.

Consider a hypothetical scenario where a rogue nation or terrorist group aims to eliminate a specific ethnic group. By analyzing the genetic markers unique to that group, scientists could use CRISPR to modify a virus or bacterium to recognize and attack only those individuals. For instance, if a specific gene variant common among the target population codes for a particular cell surface protein, the engineered pathogen could be designed to bind exclusively to that protein, initiating infection only in those who carry the genetic marker.

One potential target for such a bioweapon could be the Duffy antigen receptor, a protein on the surface of red blood cells. This receptor is present in the majority of the world's population but is notably absent in most West African individuals. By engineering a pathogen to interact specifically with the Duffy antigen, a bioweapon could theoretically spare West Africans while targeting others. Conversely, an antagonist pathogen could be designed to attack those lacking the Duffy antigen, disproportionately affecting West African populations.

Another example is the sickle cell gene, prevalent in people of African descent. While sickle cell trait offers some protection against malaria, an engineered pathogen could exploit this genetic marker to target those with the trait. Such specificity would make it difficult to detect the bioweapon's artificial origins, as the outbreak would seem to follow a natural disease pattern affecting specific genetic profiles.

In a more advanced scenario, researchers could exploit genetic variations in immune system genes, such as the HLA (human leukocyte antigen) complex. The HLA genes are highly variable and play a critical role in immune response. A pathogen engineered to bypass immune defenses based on specific HLA types could cause a devastating outbreak among individuals with those types while sparing others with different HLA profiles. This approach could be used to target ethnic groups with particular HLA distributions, making it a tool for ethnic cleansing.

Such bioweapons could be insidious because they might initially appear as natural outbreaks. Health officials would struggle to identify the true nature of the disease, delaying appropriate responses and exacerbating the impact. The pathogen's specificity could also make it challenging to develop broad-spectrum treatments or vaccines, as traditional public health measures might not be effective against a genetically targeted disease.

The ability to design bioweapons targeting specific genetic profiles using CRISPR technology represents a terrifying frontier in biological warfare. The potential for selective genocide or ethnic cleansing through such means is a grave threat that demands immediate attention from the global community. Preventing the misuse of genetic engineering for nefarious purposes will require concerted efforts in regulation, ethical standards, and international cooperation.

Beyond targeting genetic profiles, the scope of CRISPR's potential misuse extends even further into the realm of microbiomes. The human microbiome, which consists of the trillions of microorganisms living in and on our bodies, plays a crucial role in our health. CRISPR could be used to manipulate these microorganisms to produce harmful substances or to disrupt normal bodily functions. For example, altering gut bacteria to produce toxins could lead to severe gastrointestinal diseases or other health problems. This approach could be used to weaken or incapacitate enemy soldiers or civilian populations without the need for conventional weapons.

One specific example involves the gut microbiome, which is integral to digestion, immune function, and even mental health. CRISPR could be used to engineer common gut bacteria, such as *Escherichia coli* or *Bacteroides fragilis*, to produce harmful toxins. These genetically modified bacteria could be introduced into the food or water supply, leading to widespread gastrointestinal distress, severe diarrhea, and dehydration among those affected. Such a bioweapon could incapacitate soldiers, reducing their combat effectiveness and potentially leading to high mortality rates due to the secondary effects of dehydration and weakened immune responses.

Another potential target within the microbiome is the skin. The skin microbiome plays a protective role, helping to fend off pathogenic bacteria and viruses. CRISPR could be used to modify benign skin bacteria, such as *Staphylococcus epidermidis*, to express virulent factors that compromise the skin's barrier function or produce toxins that cause skin lesions and infections. This could lead to widespread dermatological issues, reducing the morale and physical readiness of troops and causing panic among civilian populations.

Respiratory microbiomes could also be manipulated. The respiratory tract contains microorganisms that help maintain respiratory health. Altering bacteria such as *Streptococcus pneumoniae* or *Haemophilus influenzae* to produce harmful agents could result in severe respiratory infections, leading to pneumonia, bronchitis, or other respiratory conditions. This would be particularly devastating in close quarters, such as military barracks or urban environments, where respiratory illnesses can spread rapidly.

Furthermore, CRISPR could be used to disrupt the oral microbiome. The mouth is home to bacteria like *Streptococcus mutans*, which play a role in oral health. Genetically engineering these bacteria to produce acids or other harmful substances could lead to rapid dental decay, gum disease, and even systemic infections if the bacteria enter the bloodstream. This could undermine the overall health and operational capability of soldiers, as dental issues can be severely debilitating.

In addition to targeting specific microbiomes, CRISPR could be used to create microorganisms that interfere with the human microbiome's overall balance, leading to dysbiosis. This imbalance can cause a wide range of health issues, including metabolic disorders, autoimmune diseases, and mental health conditions like depression and anxiety. By strategically disrupting the microbiome's equilibrium, an adversary could weaken a population over time, making them more susceptible to other illnesses and less capable of sustaining prolonged physical or mental exertion.

Such bioweapons present a unique threat because they can be introduced covertly and cause damage over a prolonged period, making detection and attribution challenging. Unlike conventional weapons, which produce immediate and visible destruction, microbiome-targeted bioweapons could silently spread through populations, causing gradual but severe health declines.

The potential use of CRISPR in biological warfare extends beyond immediate destruction to more insidious strategies, such as the induction of chronic diseases. By targeting genes associated with major health conditions like cancer, diabetes, or cardiovascular disease, CRISPR could be weaponized to introduce genetic mutations that predispose individuals to these ailments. For instance, mutations in the BRCA1 or BRCA2 genes significantly increase the risk of developing breast and ovarian cancers. A bioweapon designed to spread a CRISPR construct targeting these genes could quietly introduce these mutations into a population. Over time, the incidence of cancer would rise, straining healthcare systems, increasing mortality rates, and causing widespread fear and uncertainty.

Similarly, CRISPR could be used to induce mutations in genes linked to diabetes, such as the HNF1A gene, which is associated with maturity-onset diabetes of the young (MODY). By subtly altering these genes, a population could experience a gradual increase in diabetes cases, leading to long-term health complications and increased healthcare costs. Cardiovascular diseases could be similarly targeted by manipulating genes like LDLR, which is responsible for regulating cholesterol levels. Inducing hypercholesterolemia in a population would lead to higher rates of heart attacks and strokes, further destabilizing societal health and productivity.

Another chilling application of CRISPR in biological warfare is the potential to weaken the immune system by knocking out genes critical for immune function. The gene encoding the major histocompatibility complex (MHC), for example, plays a crucial role in the immune system's ability to recognize and fight infections. Disabling this gene would severely compromise the immune system, making individuals more susceptible to infections and diseases. A bioweapon designed to spread a CRISPR construct that disrupts MHC genes could be particularly devastating.

Imagine a scenario where a CRISPR-based bioweapon is released in conjunction with a highly infectious pathogen, such as a modified influenza virus. The initial CRISPR construct would knock out essential immune system genes, leaving the population vulnerable. The subsequent release of the pathogen would then spread unchecked, causing widespread illness and death. Such a strategy could cripple healthcare systems, as doctors and hospitals struggle to cope with an overwhelming number of patients who are unable to mount an effective immune response.

Imagine a world where today's fiction becomes tomorrow's reality. As you delve into the following story, envision a scenario where the events described could be the leading headline on your local news or the top story on your favorite news channel. Consider the profound implications and chilling possibilities that could unfold, making this narrative not just a tale of the future but a potential warning for the present.

In the year 2045, tensions between Pakistan and India reached a breaking point, leading to a covert operation by Pakistan to deploy a bioweapon using advanced CRISPR technology. The target was India's densely populated cities, and the plan was executed with chilling precision.

The operation began with agents releasing the CRISPR construct via aerosols from a passing train at several major train stations during peak hours. As commuters boarded and disembarked, the invisible genetic modifier dispersed into the air, silently infiltrating their systems. The construct was designed to knock out critical immune system genes, rendering the immune systems of those exposed ineffective.

Unaware of the insidious assault, millions of people went about their daily lives as the CRISPR construct spread rapidly through the urban population. Within a few days, the second phase of the operation commenced. Pakistan deployed agents disguised as street vendors, maintenance workers, and even as medical personnel, strategically placing them in crowded areas such as markets, shopping malls, and public transportation hubs. These agents carried inconspicuous devices designed to release a genetically modified influenza virus in the form of a fine mist.

At predetermined times, the agents activated these devices, ensuring maximum exposure during peak activity periods. The virus, now faced with a population whose immune defenses had been sabotaged by the CRISPR construct, spread unchecked, leading to widespread illness and death. Public transportation systems, including buses and metro trains, became prime vectors for the pathogen, as the confined spaces facilitated rapid transmission among passengers.

The impact on India's healthcare system was immediate and catastrophic. Hospitals were inundated with patients exhibiting severe flu symptoms, but standard treatments proved futile. Medical supplies ran out within days, and healthcare workers, also vulnerable to the virus, succumbed in alarming numbers. Makeshift quarantine zones were established in public spaces like stadiums and schools, but these quickly became overwhelmed as the death toll mounted.

Indian scientists, realizing the nature of the attack, worked around the clock to develop a countermeasure. They aimed to create a new CRISPR construct to repair the damaged immune genes, but the complexity of the task and the rapid spread of the virus meant that progress was slow. For countless victims, any solution would come too late.

This unprecedented bioweapon attack showcased the terrifying potential of genetic engineering when used maliciously. The event underscored the urgent need for international regulations on biotechnology and heightened preparedness for bioethical threats. As the world watched in horror, the attack served as a grim reminder of the delicate balance between scientific progress and its potential for misuse, leaving deep scars on the global consciousness and prompting a reevaluation of ethical standards in genetic research.

The devastating attack led to swift and severe repercussions. With overwhelming evidence pointing to Pakistan as the orchestrator of the bioweapon strike, India declared war. The already strained relations between the two nuclear-armed neighbors plunged into outright conflict. The world held its breath as the potential for escalation threatened global stability. The international community *urgently called for peace, fearing the catastrophic consequences of a full-scale war between India and Pakistan, but the scars left by the CRISPR attack drove the conflict forward into an uncertain and dangerous future.*

The implications of these CRISPR-based bioweapons are profound. They could destabilize societies not through immediate, visible destruction, but through a slow, insidious erosion of public health. The economic impact would be enormous, with skyrocketing healthcare costs, reduced workforce productivity, and long-term suffering for those affected. Additionally, the difficulty in tracing the source of such an attack would complicate international responses and accountability, potentially leading to geopolitical tensions and conflicts.

The potential uses of CRISPR in biological warfare include inducing chronic diseases and weakening immune systems. These strategies could have devastating long-term effects on populations, healthcare systems, and economies. However, the implications of CRISPR-based bioweapons extend even further, potentially targeting the very foundation of a nation's sustenance.

CRISPR technology, while holding immense potential for beneficial applications, also poses serious risks if misused as a bioweapon. One particularly alarming scenario involves targeting agricultural crops and livestock, which could lead to devastating food shortages and economic instability. By using CRISPR to modify plant or animal pathogens to be more virulent or resistant to existing treatments, an adversary could effectively destroy food supplies, causing widespread famine and destabilizing nations.

For instance, consider staple crops like wheat, rice, and corn, which are crucial for the global food supply. If CRISPR were used to enhance the virulence of pathogens like Puccinia graminis, the fungus responsible for wheat stem rust, or Magnaporthe oryzae, which causes rice blast disease, the resulting strains could be devastating. These genetically modified pathogens could overcome the plants' natural defenses and existing fungicides, leading to rapid and uncontrollable outbreaks. Fields of wheat or rice could be wiped out in a single growing season, leading to immediate shortages.

Similarly, livestock could be targeted by modifying viruses such as the African swine fever virus (ASFV) or the foot-and-mouth disease virus (FMDV). ASFV is already a significant threat to pig populations, and using CRISPR to create a more virulent or treatment-resistant strain could devastate the pork industry. The resulting mass culling of infected animals would not only lead to a shortage of pork but also have ripple effects throughout the economy, affecting feed suppliers, processors, and retailers. FMDV, which affects cattle, pigs, sheep, and goats, could be similarly enhanced to resist vaccines and treatments, leading to widespread livestock losses and further compounding food shortages.

The economic impact of such bioweapons would be profound. Agriculture is a major sector in many economies, and the sudden loss of crops and livestock would cause prices to skyrocket, making food unaffordable for many. This would lead to increased poverty and hunger, and potentially spark social unrest as people compete for scarce resources. Farmers and agricultural businesses would face ruin, and the costs of managing and attempting to recover from such outbreaks would strain national budgets and resources.

The broader societal impacts would also be severe. In countries heavily reliant on agriculture, such attacks could undermine trust in the government's ability to protect food supplies, leading to loss of public confidence and potential political instability. International trade would be affected as nations impose bans and restrictions to prevent the spread of modified pathogens, disrupting global food markets and leading to further shortages and price volatility.

Moreover, the environmental consequences could be dire. The use of more virulent pathogens could lead to the collapse of ecosystems that rely on certain crops or livestock. The loss of biodiversity, as a result, would have long-term effects on the environment, reducing the resilience of ecosystems to other stresses like climate change.

This type of bioweapon, by targeting the foundational elements of food security, could cripple a nation's ability to sustain its population and maintain order. The potential for CRISPR to be misused in this manner underscores the urgent need for robust international regulations and biosecurity measures. Governments and global organizations must work together to monitor and control the use of genetic editing technologies, ensuring they are used responsibly and safeguarding against their potential weaponization.

The potential use of CRISPR for creating biological weapons highlights a critical gap in current international regulations. The Biological Weapons Convention (BWC), which prohibits the development and use of biological weapons, does not explicitly address the use of genetic modifications for military purposes. This regulatory gap could be exploited by nations or non-state actors to develop and deploy genetic bioweapons without clear legal repercussions.

As the capabilities of CRISPR and other gene-editing technologies continue to advance, there is an urgent need for the international community to reconsider existing treaties and international laws governing warfare. The BWC and other frameworks must be updated to explicitly prohibit the use of genetic modifications for creating biological weapons. Additionally, new mechanisms for monitoring and enforcement must be established to ensure compliance and to prevent the misuse of these powerful technologies.

To address these challenges, the international community could consider several approaches. Amending the BWC to explicitly include genetic modifications and CRISPR technology within its scope would ensure that any development of bioweapons using these methods is clearly prohibited. Developing robust verification mechanisms to monitor compliance with the BWC, including regular inspections of research facilities and increased transparency in genetic research, is also necessary. Fostering international collaboration and information sharing to detect and prevent the misuse of CRISPR technology could involve creating a global database of genetic research and enhancing communication between scientific communities and regulatory bodies. Developing and promoting ethical guidelines for the use of CRISPR and other gene-editing technologies in both civilian and military contexts should emphasize the importance of preventing misuse and ensuring that genetic research is conducted responsibly.

While CRISPR technology holds great promise for advancing human health and capabilities, its potential use in creating biological weapons poses significant risks. The international community must take proactive steps to close regulatory gaps and establish new frameworks to govern the use of gene-editing technologies in warfare. By doing so, we can help ensure that these powerful tools are used for the benefit of humanity and not as instruments of destruction.

In this rapidly evolving landscape, it is crucial for policymakers, military leaders, and international organizations to engage in dialogue and develop strategies to address the potential risks and benefits of CRISPR-modified soldiers and organisms. The goal should be to ensure that genetic enhancements are used responsibly and ethically, preventing their misuse and mitigating the potential for conflict.

Across the globe, numerous countries have embarked on researching the military applications of CRISPR technology. While some nations are just beginning to explore its potential, others are actively racing to leverage these genetic advancements to gain a strategic edge over their international rivals. Countries like the United States and China have already invested heavily in CRISPR research, not only for medical advancements but also for potential defense applications. In 2018, the U.S. Defense Advanced Research Projects Agency (DARPA) announced significant funding for gene-editing technologies, underscoring their potential to protect against biological threats and enhance soldier performance. This initiative, part of DARPA's broader effort to maintain technological superiority and national security, aims to explore the dual-use capabilities of CRISPR in both defensive and offensive military contexts.

DARPA's investment in CRISPR technology is multifaceted. Programs like Safe Genes focus on developing robust methods to control gene-editing activities, ensuring that such technologies can be used safely and effectively. These initiatives include creating genetic "off-switches" to prevent unintended or harmful genetic modifications, which are crucial for the responsible deployment of gene-editing tools. Additionally, DARPA's research into CRISPR includes projects aimed at enhancing the physical and cognitive abilities of soldiers. By potentially enabling faster recovery from injuries, increased resistance to environmental stresses, and improved mental acuity, CRISPR could fundamentally transform the capabilities of military personnel.

Similarly, China's investments in biotechnology have been substantial, reflecting a strategic interest in leveraging CRISPR for both civilian and military purposes. Reports suggest that the Chinese military is actively exploring the use of gene-editing technologies to create enhanced combatants. This includes research into increasing physical strength, endurance, and resilience among soldiers. China's approach to CRISPR research is part of a broader national strategy to achieve technological dominance in key scientific fields, including biotechnology.

In 2017, Chinese scientists made headlines with the successful editing of human embryos to remove a genetic mutation that causes a potentially fatal blood disorder. While the primary focus of this research was medical, the techniques and insights gained have clear implications for military applications. The ability to edit genes with precision opens up the possibility of enhancing human performance in ways that could provide a significant advantage on the battlefield.

The strategic investments by both the United States and China in CRISPR technology highlight the emerging arms race in genetic engineering. This competition is not only about achieving medical breakthroughs but also about gaining a strategic edge in future conflicts. As both nations continue to push the boundaries of CRISPR research, the ethical and security implications of such advancements will become increasingly prominent. The potential to create a new class of enhanced soldiers raises profound questions about the future of warfare and the role of genetic technology in shaping global power dynamics.

Consider the NBC News article from 2020 titled "China has done human testing to create biologically enhanced super soldiers, says top U.S. official". U.S. intelligence has revealed that China has conducted "human testing" on members of the People's Liberation Army with the aim of developing soldiers with "biologically enhanced capabilities," according to John Ratcliffe, the U.S. director of national intelligence. This assertion was made in a Wall Street Journal op-ed, emphasizing China as a significant national security threat to the U.S.

Ratcliffe's claim, suggesting a pursuit similar to fictional "super soldiers," was not elaborated upon by his office or the CIA. Last year, American scholars Elsa Kania and Wilson VornDick published a paper indicating China's interest in applying biotechnology, particularly CRISPR, to enhance human performance on the battlefield. CRISPR, a gene-editing tool, is ethically controversial when used to enhance healthy individuals' capabilities. The scholars highlighted China's view of biotechnology as a future strategic asset in military affairs. They cited a 2017 statement by a Chinese general on the revolutionary impact of integrating biotechnology with other advanced fields on warfare.

VornDick expressed concerns about the unforeseen consequences of genetic modifications. The Chinese government did not comment on these allegations. Ratcliffe's broader message was that China poses the greatest current threat to American and global security, urging President-elect Joe Biden to acknowledge this threat.

These developments underscore the need for international dialogue and regulation to address the ethical and security challenges posed by genetic engineering. As countries like the United States and China forge ahead with their CRISPR research, the international community must grapple with the implications of these powerful technologies and work towards frameworks that ensure their responsible use.

Similarly to China, Russia has also shown significant interest in genetic technologies for military applications, signaling its intent to remain at the forefront of this emerging field. In an interview from 2017, Russia's President Vladimir Putin warned that "Humanity could soon create something worse than a nuclear bomb. One may imagine that a man can create a man with some given characteristics, not only theoretically but also practically. He can be a genius mathematician, a brilliant musician or a soldier, a man who can fight without fear, compassion, regret or pain."

Despite Putin's superficial concern, the Russian military is reportedly implementing "genetic passports" for its personnel, a groundbreaking initiative designed to assess genetic predispositions and optimize role assignments based on an individual's genetic makeup. This approach aims to leverage genetic insights to place soldiers in roles where they can perform most effectively, thereby enhancing overall military efficiency and capability.

The "genetic passport" program is part of a broader, ambitious effort to boost human performance through cutting-edge genetic research. By analyzing the genetic profiles of its soldiers, the Russian military hopes to identify traits such as physical endurance, stress tolerance, and cognitive abilities, which can then be matched to specific military tasks. For instance, individuals with genetic markers for high stress resilience might be assigned to high-pressure environments like special operations or intelligence units, while those with superior physical attributes could be directed towards combat roles requiring significant physical exertion.

Russian President Vladimir Putin has made it clear that genetic research is a strategic priority for national defense. Through a series of decrees, he has mandated the integration of genetic profiling into the country's defense strategies, underscoring the critical importance of genetic research in maintaining and advancing Russia's military prowess. These decrees highlight the potential of genetic technologies not only to enhance individual soldier performance but also to contribute to broader military capabilities.

In addition to genetic passports, Russia is reportedly investing heavily in various other aspects of genetic research, positioning itself at the forefront of a new kind of biological arms race. This includes exploring advanced gene-editing technologies like CRISPR with the aim of potentially developing enhanced soldiers. These soldiers could possess augmented physical and cognitive abilities far beyond the current capabilities of ordinary humans. Russian military and scientific communities are not only focusing on immediate applications but are also delving into long-term genetic enhancements that could provide a strategic edge in future conflicts.

The scope of this research is expansive, covering a range of genetic modifications that could revolutionize military effectiveness. For instance, there is a significant focus on the genetic basis of rapid wound healing. By identifying and enhancing genes responsible for tissue regeneration, Russian scientists aim to create soldiers who can recover from injuries much faster than normal, reducing downtime and maintaining combat readiness.

For a country like Russia, which encompasses a vast expanse of frigid and often inhospitable territory, enhancing resistance to extreme environmental conditions through genetic research holds significant strategic and operational benefits. The ability to modify soldiers to better withstand harsh climates would be particularly advantageous in several ways.

Russia has extensive interests in the Arctic region, which is rich in natural resources and holds significant geopolitical importance. The ability to deploy soldiers who are genetically enhanced to resist extreme cold would allow for sustained military operations in these icy conditions. Enhanced resistance to cold would prevent common cold-related injuries such as frostbite and hypothermia, maintaining soldiers' health and readiness.

Modifications that enable soldiers to better regulate their body temperature and metabolic processes in extreme cold would enhance survival rates. Soldiers would be less likely to succumb to environmental stresses that typically impair performance. This means they could stay active and alert for longer periods, perform their duties more effectively, and recover more quickly from exposure to harsh conditions.

Russia's geography includes vast regions with severe climates, from Siberia to the mountainous areas of the Far East. Soldiers with enhanced resistance to these environments could conduct operations more efficiently, whether it's patrolling borders, responding to natural disasters, or engaging in combat scenarios. This adaptability would provide a strategic advantage, allowing Russian military forces to operate in areas where adversaries might struggle.

The ability to operate effectively in harsh climates would extend the deployment capabilities of Russian military units. Soldiers could be stationed in remote, strategic locations for longer durations without the need for frequent rotations or extensive logistical support to mitigate the effects of the environment. This would enhance Russia's ability to maintain a constant and robust military presence in critical areas.

In addition to cold environments, modifications that enhance resistance to a range of extreme conditions, including heat, would make Russian forces more versatile. This would be beneficial for operations in diverse climates, such as desert regions in the Middle East where Russia has political and military interests. Soldiers who can adapt to various environmental extremes would be an asset in global deployments.

Enhancing soldiers' ability to withstand extreme conditions could reduce the need for specialized equipment and clothing designed to protect against harsh climates. This would result in economic savings and logistical efficiencies, as the burden of transporting and maintaining such equipment would be lessened. Resources could then be allocated to other critical areas, enhancing overall military effectiveness.

One historical example of how cold climates affected a war's outcome is Napoleon's invasion of Russia in 1812. The harsh Russian winter played a crucial role in the French army's defeat and had a significant impact on the course of the Napoleonic Wars.

In June 1812, Napoleon Bonaparte launched his Grande Armée, consisting of over 600,000 soldiers, into Russia with the objective of forcing Tsar Alexander I to remain in the Continental Blockade against Britain. Initially, Napoleon's army advanced quickly, but as they penetrated deeper into Russian territory, they faced severe logistical challenges. The vast distances and scorched-earth tactics employed by the Russians, who burned their own villages and crops to deny resources to the French, began to strain Napoleon's supply lines.

The French forces captured Moscow in September 1812, but instead of suing for peace, the Russians retreated and left the city largely abandoned. With winter approaching and no decisive victory in sight, Napoleon's army faced a dire situation. By October, Napoleon decided to retreat from Moscow.

The retreat proved disastrous. The early onset of a brutal winter brought temperatures as low as -30 degrees Celsius (-22 degrees Fahrenheit). The French soldiers, unprepared for such extreme cold, suffered from frostbite, starvation, and disease. The cold immobilized troops and horses, making movement difficult and slow. The Russian army and partisans constantly harassed the retreating French forces, further compounding their misery.

By the time the remnants of the Grande Armée crossed the Berezina River in late November, only about 27,000 soldiers remained combat-effective. The retreat from Russia decimated

Napoleon's army, with the vast majority of his troops killed, captured, or incapacitated by the cold and other hardships.

The failure of the Russian campaign marked a turning point in the Napoleonic Wars. It significantly weakened Napoleon's military power and emboldened his enemies across Europe, leading to a coalition that would eventually defeat him. The Russian winter, with its extreme cold and logistical challenges, played a pivotal role in the collapse of Napoleon's campaign and the eventual downfall of his empire.

Russian soldiers who are genetically enhanced to better cope with extreme environments would likely experience improved psychological resilience. Knowing they can withstand harsh conditions would boost their confidence and morale, leading to more cohesive and effective units. This psychological edge could be crucial in maintaining high performance levels during prolonged and demanding operations.

In addition to the United States, China, and Russia, other countries are also exploring the use of CRISPR and other genetic technologies for military applications. The United Kingdom, for instance, has shown significant interest in the realm of genetic engineering for defense purposes. The UK government has been actively investing in genetic defense research, as highlighted in their comprehensive national defense reviews. These reviews consistently emphasize the strategic importance of advancements in genetic engineering for maintaining national security and defense capabilities. They outline a vision where genetic modifications could enhance the physical and cognitive abilities of military personnel, offering a significant tactical advantage.

Central to these efforts is the UK's Advanced Research and Invention Agency (Aria), a new initiative modeled after the United States' DARPA. Aria is poised to spearhead groundbreaking research and innovation in genome technology, specifically tailored for defense applications. With a mandate to explore and develop cutting-edge technologies, Aria is expected to push the boundaries of what is possible with genetic engineering. This includes the potential development of genetically enhanced soldiers, improved resilience against biological threats, and novel biotechnological solutions to contemporary military challenges.

However, Aria's operations have sparked a degree of controversy. Unlike many other government agencies, Aria operates with a high degree of autonomy and significantly less public accountability. This lack of transparency has raised concerns among ethicists, policymakers, and the general public. The fear is that, without rigorous ethical oversight, the pursuit of advanced genetic technologies could lead to ethical transgressions and unintended consequences. Critics argue that the potential for misuse or reckless experimentation increases when oversight mechanisms are weak or absent.

Moreover, the secretive nature of Aria's projects has fueled speculation and concern. Given the sensitive nature of genetic research, particularly in the context of military applications, the call for greater transparency and ethical regulation is growing louder. The debate centers around ensuring that advancements in genetic engineering do not come at the expense of ethical standards or public trust. As the UK continues to advance its genetic defense research, balancing innovation with ethical responsibility will be crucial in navigating the complex landscape of modern military technology.

France is another country that has been actively exploring the potential of genetic technologies for military purposes. The French military's interest in genetic enhancement is evidenced by the recent approval from their military ethics committee to conduct research into enhancing soldiers through genetic modifications. This landmark decision reflects a significant shift in military strategy, highlighting a willingness to delve into and potentially implement genetic enhancements that could revolutionize the physical and cognitive abilities of military personnel.

The French military's move towards genetic research is not just a theoretical exploration but a pragmatic step towards future-proofing their armed forces. By investigating genetic modifications, France aims to develop soldiers who possess superior strength, enhanced endurance, and quicker recovery times from injuries.

France's exploration into genetic technologies is part of a broader trend among leading military powers to integrate advanced biotechnologies into their defense strategies. This initiative aligns with global efforts to leverage cutting-edge science to enhance national security and maintain military superiority. The ethical approval by the French military's ethics committee underscores a recognition of the profound implications such research entails, balancing the pursuit of technological advancements with considerations of moral and ethical responsibility.

This move also signals France's commitment to staying at the forefront of military innovation, ensuring that its armed forces are not only equipped with the latest conventional weapons but are also enhanced at a biological level. The implications of such advancements are vast, potentially redefining the standards of military excellence and reshaping the future of warfare. As research progresses, the French military's approach to genetic modifications will likely serve as a bellwether for other nations considering similar enhancements, setting precedents in both technological capabilities and ethical frameworks.

The potential for an arms race in genetic technology raises numerous ethical and strategic concerns. Unlike traditional weapons, which are subject to various international treaties and regulations, genetic enhancements present a new frontier with relatively few existing controls. The Biological Weapons Convention (BWC), which outlaws biological and toxin weapons, does not specifically address the nuances of genetic modification for military purposes. This regulatory gap could lead to unchecked advancements, as nations may prioritize national security over international cooperation and ethical considerations.

Moreover, the strategic advantage gained by nations that successfully develop CRISPR-enhanced soldiers could be profound. A military force with genetically modified troops could outperform conventional soldiers, shifting the power dynamics in conflicts. This advantage could pressure other nations to pursue similar technologies, sparking a cycle of competitive enhancements reminiscent of the nuclear arms buildup.

The implications of a genetic arms race extend beyond the battlefield. The societal and moral ramifications are immense, as the line between human and machine becomes increasingly blurred. The pursuit of genetic superiority could lead to a new era of eugenics, where the definition of human worth and potential is dictated by genetic enhancements. This dystopian outcome, while speculative, underscores the urgent need for international dialogue and regulatory frameworks to govern the use of CRISPR in military contexts.

As the technology of CRISPR soldiers becomes more prevalent, it is likely that an increasing number of countries, including those with more unscrupulous rulers, will begin to use and potentially abuse this powerful technology. For instance, North Korea under Kim Jong-un, known for its aggressive military stance and disregard for international norms, could exploit CRISPR to create a new breed of super-soldiers, enhancing their physical and cognitive abilities without ethical oversight. Similarly, China under Xi Jinping has demonstrated a strong interest in advancing military technologies and may leverage genetic modifications to gain a strategic edge. In such hands, CRISPR technology could be used to create enhanced troops, raising the specter of human rights abuses, forced genetic modifications, and the deployment of these soldiers in aggressive and oppressive ways. This proliferation poses significant risks, as it could lead to an arms race in genetic enhancements, destabilizing international security and exacerbating global tensions. The lack of stringent international regulations and oversight mechanisms further compounds these dangers, making it imperative for the global community to address the ethical and legal challenges posed by the widespread adoption of CRISPR-based military enhancements.

While the public domain is aware of some nations' interest in CRISPR technology for military purposes, it is widely believed that many countries have conducted far more extensive research and testing than they openly disclose. These efforts are often shrouded in secrecy, as the strategic advantages offered by genetic modifications are considered highly sensitive military information. Governments are keen to keep their advancements under wraps to maintain an edge over potential adversaries.

In this clandestine race, countries invest heavily in cutting-edge genetic research, working in secure facilities away from prying eyes. These secret programs explore a range of applications, from enhancing the physical and cognitive abilities of soldiers to developing more resilient crops and livestock for military logistics. By keeping these developments confidential, nations aim to surprise their rivals with capabilities that are not yet widely understood or countered.

Moreover, the implications of revealing such research are significant. Publicly acknowledging advanced genetic modification programs could provoke international condemnation and lead to stringent regulatory measures that might stifle progress. Hence, nations choose to veil their CRISPR endeavors in secrecy, sharing information only on a need-to-know basis within tightly controlled circles of military and scientific personnel.

As we navigate this uncharted territory, the question remains: can the global community find a balance between leveraging the benefits of CRISPR technology and preventing its misuse? The stakes are high, and the future of genetic enhancement in warfare will depend on the collective actions of nations, scientists, and policymakers.

Chapter 6.

The Science of Super-Soldiers

The process of creating genetically modified soldiers using CRISPR technology presents numerous technical challenges that need to be addressed. While the potential for enhanced physical and cognitive abilities is promising, the complexity of human genetics and the limitations of current technology pose significant hurdles.

One of the primary challenges is the precise editing of the human genome. Although CRISPR-Cas9 has revolutionized the field of genetic engineering with its ability to target specific DNA sequences, it is not without its flaws. Off-target effects, where CRISPR inadvertently edits unintended parts of the genome, remain a major concern. These unintended modifications can lead to unpredictable consequences, potentially causing harmful mutations or disrupting essential genes. A study published in the journal *Nature Methods* underscored the potential dangers associated with CRISPR technology, revealing that it can inadvertently introduce numerous off-target mutations. These unintended genetic changes occur when CRISPR-Cas9, while editing the target gene, also makes alterations in other parts of the genome. Such off-target mutations could have unpredictable and potentially serious implications for the health and safety of genetically modified soldiers.

The findings from this study are particularly concerning in a military context, where precision and reliability are paramount. Off-target effects could result in a range of unintended consequences, from minor health issues to severe, life-threatening conditions. For example, an off-target mutation might activate oncogenes, leading to an increased risk of cancer, or disrupt genes critical for immune function, making soldiers more susceptible to infections.

Moreover, the cumulative effects of multiple off-target mutations could compromise the overall physical and mental performance of genetically modified soldiers. This could negate the intended benefits of genetic enhancements, leading to soldiers who are not only less effective but also at greater risk of long-term health problems. The unpredictability of these genetic changes also poses a significant challenge for medical monitoring and treatment, as it may be difficult to identify and manage the diverse range of potential side effects.

These risks raise serious ethical and practical questions about the deployment of CRISPR technology in military applications. While the promise of creating super-soldiers with enhanced abilities is alluring, the potential for harmful genetic side effects cannot be ignored. The study from *Nature Methods* serves as a critical reminder of the need for rigorous testing and ethical considerations in the development of gene-editing technologies, particularly when applied to humans.

Another significant challenge is the delivery of CRISPR components into the human body. Efficient and targeted delivery systems are crucial for the success of genetic modifications. Viral vectors, such as adenoviruses, are commonly used to deliver CRISPR components into cells, but they carry risks of immune responses and insertional mutagenesis. Non-viral delivery methods, like lipid nanoparticles, are being explored, but they currently lack the efficiency and specificity needed for widespread clinical application. Researchers are continuously developing and testing new delivery methods to overcome these obstacles, but a universally effective and safe system has yet to be realized.

Furthermore, the complexity of human traits, especially those related to physical and cognitive enhancements, adds another layer of difficulty. Most traits are polygenic, meaning they are influenced by multiple genes, and the interactions between these genes are not fully understood. For instance, enhancing muscle strength might involve editing several genes related to muscle fiber composition, metabolism, and growth factors. Each of these genes can have pleiotropic effects, impacting various biological pathways and potentially leading to unforeseen side effects. The intricate interplay of genetic networks means that a modification intended to enhance one trait could inadvertently impair another, making the task of creating a well-rounded genetically modified soldier exceedingly complex.

Ethical considerations also intersect with technical challenges. The potential for germline editing, where genetic modifications are passed onto future generations, raises significant ethical and safety concerns. Although the focus of military applications might be on somatic cell editing, which affects only the individual and not their offspring, the long-term effects of such modifications remain unknown. The ethical implications of creating enhanced individuals who might have an unfair advantage over others, and the potential societal impact, must be carefully weighed against the perceived benefits.

One of the most significant advancements in CRISPR technology is the development of high-fidelity Cas9 variants. These engineered enzymes have reduced off-target activity, making genome editing more precise and safer. According to a 2018 study published in *Nature*, these high-fidelity Cas9 variants, such as SpCas9-HF1 and eSpCas9, exhibit significantly fewer off-target effects compared to their predecessors, thereby increasing the reliability of CRISPR as a tool for genetic modification.

The complexity of the human genome poses additional hurdles. The human genome contains vast regions of non-coding

DNA, repetitive sequences, and structural variations that can complicate the editing process. Ensuring that the CRISPR-Cas9 system targets the correct location without disrupting other important genetic elements remains a significant challenge. A study published in *Genome Biology* in 2019 highlighted these complexities, noting that even with high-fidelity variants, the potential for unintended genomic alterations cannot be entirely eliminated.

Another critical limitation is the current inability to control the repair mechanisms that follow the DNA cut made by Cas9. After Cas9 makes a double-strand break, the cell's natural repair processes take over. The two primary pathways, non-homologous end joining (NHEJ) and homology-directed repair (HDR), can lead to unpredictable outcomes. NHEJ, which is more common, often results in insertions or deletions that can introduce unintended mutations. HDR, while more precise, requires a template for repair and is less efficient in most cells. Researchers are actively working on improving these repair pathways to enhance the precision and predictability of CRISPR-induced edits.

As we stand on the cusp of these technological advancements, it is essential to recognize both the extraordinary potential and the inherent risks of CRISPR technology. The path forward will require careful consideration of ethical, technical, and safety concerns, ensuring that the quest for genetic perfection does not overshadow the fundamental principles of responsible science.

Moving beyond the intricacies of CRISPR's precision and limitations, it is crucial to understand how these technological advancements are being applied in various domains, particularly in the realm of military enhancements. The intersection of CRISPR and military aspirations presents a complex landscape where scientific innovation meets ethical and strategic considerations, shaping the future of warfare in unprecedented ways.

On the technical front, George Church, a prominent geneticist at Harvard University, has been exploring the extensive possibilities of genetic modifications. Church's work includes attempts to enhance human resilience to diseases and aging, which could directly translate to more robust soldiers. However, he cautions against the premature application of such technologies. Church notes that the long-term effects of genetic modifications are still largely unknown, and off-target effects—unintended changes to the genome—remain a significant risk. These unintended alterations could have unforeseen health consequences, undermining the very enhancements CRISPR aims to achieve.

Biotechnologist Feng Zhang, another pivotal figure in the development of CRISPR technology, embodies a blend of cautious optimism and scientific rigor in his approach to gene editing. Recognizing the profound implications of CRISPR's potential to create "super-soldiers" with enhanced physical and cognitive abilities, Zhang maintains a balanced perspective that emphasizes both the transformative possibilities and the critical need for ethical oversight.

At the forefront of CRISPR research, Zhang's work at the Broad Institute focuses on addressing the technical challenges that accompany this powerful tool. He is dedicated to refining CRISPR technology to improve its precision and significantly reduce off-target effects, which are unintended changes to the genome that can result from the editing process. These efforts are not merely academic; they are essential steps towards ensuring the safety and efficacy of any potential applications, particularly in the highly sensitive context of military use.

Zhang's meticulous research involves developing more accurate and reliable CRISPR systems, such as CRISPR-Cas12 and CRISPR-Cas13, which offer different mechanisms of action and potentially greater specificity. By advancing these technologies, Zhang aims to create a toolkit that can be tailored to various genetic targets with minimal risk of unintended consequences. This precision is crucial for any future applications where genetic modifications must be safe, effective, and reversible if necessary.

Zhang's cautious optimism reflects a commitment to advancing scientific knowledge while safeguarding human values. His work exemplifies the dual path of innovation and responsibility, setting a high standard for how groundbreaking technologies should be developed and applied. As CRISPR continues to evolve, Zhang's contributions will undoubtedly play a crucial role in shaping a future where genetic engineering can be safely and ethically integrated into various aspects of human life, including the complex and controversial domain of military enhancement.

Another pressing concern raised by experts is the potential for genetic modifications to be inherited by future generations. Marcy Darnovsky, executive director of the Center for Genetics and Society, emphasizes the profound ethical and safety concerns associated with germline modifications—changes to the genome that can be passed down to offspring. This issue is particularly troubling when considering the military's use of genetic engineering, as the implications extend far beyond the individual soldiers being enhanced.

Germline editing, unlike somatic editing which alters genes in a non-heritable way, affects the very blueprint of human biology. While somatic editing could be confined to individual soldiers, targeting specific traits like enhanced muscle strength or faster healing without impacting their descendants, germline editing introduces permanent changes to the genetic code. This means any modifications made could propagate through future generations, potentially altering the course of human evolution.

Imagine a future where a rogue nation, driven by a desire to dominate the global stage, secretly embarks on an ambitious genetic program to create the ultimate super soldiers. Unlike conventional enhancements, which modify only the individual without affecting their descendants, this nation opts for the more radical approach of germline editing. This involves altering the very blueprint of human biology, introducing permanent changes that will be passed down through generations.

The first generation of these genetically modified soldiers emerges, displaying extraordinary physical prowess, rapid healing abilities, and superior cognitive functions. They excel in every military operation, outperforming adversaries with ease. The world watches in awe and fear as these super soldiers reshape the battlefield, giving their nation an unprecedented strategic advantage.

However, the true implications of this germline editing begin to manifest as these soldiers start to have families. The genetic modifications, designed for combat efficiency, now propagate through their offspring. These children, born with enhanced abilities, face a world unprepared for their existence. Schools struggle to accommodate their advanced physical and mental capabilities, and social integration becomes increasingly challenging.

As generations pass, the modifications continue to spread, no longer confined to the military elite. The nation's population starts to exhibit a widening genetic divide. Those with the enhanced genes thrive, while those without them are left at a significant disadvantage, fostering a new form of genetic inequality. Social tensions escalate as the unenhanced populace demands equal opportunities and rights, leading to widespread unrest and instability.

Moreover, unforeseen genetic side effects begin to appear. While the original enhancements were intended for battlefield advantages, they inadvertently disrupt natural biological processes. Increased susceptibility to certain diseases, unanticipated psychological effects, and reduced genetic diversity result in a population increasingly vulnerable to new health crises. The genetic modifications, once seen as a miraculous advancement, now threaten the very fabric of society.

Internationally, other nations scramble to respond to this new genetic arms race. In a bid to keep up, they launch their own germline editing programs, further propagating the cycle of genetic modification. The global genetic landscape shifts dramatically, with humanity's natural evolutionary path irrevocably altered.

In this grim scenario, the reckless use of germline editing for military superiority reveals its dark side. The initial promise of creating super soldiers morphs into a dystopian reality where genetic inequality, unforeseen health consequences, and social instability threaten the future of humanity. The world learns a harsh lesson: tampering with the fundamental aspects of human biology carries risks that extend far beyond the battlefield, potentially altering the course of human evolution in ways that cannot be undone.

Germline modifications could lead to a new form of genetic inequality, where certain traits deemed desirable are perpetuated within select groups, while others are left behind. This could exacerbate existing social divides and lead to new forms of discrimination based on genetic "purity" or enhancements. Furthermore, the long-term effects of such modifications are unknown. Unintended genetic changes could manifest generations later, potentially introducing new diseases or vulnerabilities into the human gene pool.

Safety concerns are equally significant. The complexity of human genetics means that even well-intentioned modifications could have unforeseen consequences. For instance, a gene edit intended to enhance physical strength might inadvertently increase susceptibility to other health issues. The lack of comprehensive understanding of gene interactions and epigenetics—the study of changes in organisms caused by modification of gene expression rather than alteration of the genetic code itself—adds another layer of risk.

Moreover, the potential for misuse of germline editing in a military context casts a sinister shadow over ethical considerations. The relentless drive to create super-soldiers could lead to reckless, hasty, and poorly regulated experiments, where the pursuit of power outweighs moral constraints. The prospect of engineering future generations of soldiers with predetermined traits for strength, intelligence, or unyielding obedience is not just a scientific endeavor; it is a harrowing step toward a dystopian reality. This dark ambition threatens to strip away the very essence of human agency and autonomy, turning soldiers into mere tools of warfare, their genetic fate dictated by the cold calculus of military strategy. The implications are chilling, suggesting a future where the lines between humanity and machine are blurred, and where the sanctity of human life is compromised for the sake of creating an unstoppable force on the battlefield.

While somatic editing offers a more controlled and contained approach to enhancing individual soldiers, germline editing opens Pandora's box of ethical, safety, and social issues. As we stand on the brink of these technological advancements, it is crucial to establish stringent regulations and ethical guidelines to prevent irreversible changes to the human genome that could impact countless generations to come.

Collectively, these expert opinions reveal a technology teetering on the edge of a dystopian future, where the allure of unprecedented power is shadowed by grave challenges and ethical quandaries. The scientific community's call for a cautious, measured approach to the application of CRISPR, especially within the perilous and high-stakes domain of military enhancement, often clashes with the relentless pressures exerted by the military-industrial complex. This powerful sector, driven by profit and the quest for dominance, has little patience for ethical debates or the slow march of cautious scientific progress.

In this grim landscape, the voices advocating for restraint and ethical oversight risk being drowned out by the clamor for progress and power. The pursuit of genetic modifications for military use stands on a knife-edge, where the potential for creating super-soldiers could overshadow the profound implications for humanity's future. The relentless drive of the military-industrial complex to harness CRISPR's capabilities underscores a stark reality: the race for genetic enhancement may well come at an incalculable cost, eroding the very foundations of ethical science and social order.

As we delve deeper into the tactical advantages and battlefield realities of genetically modified soldiers, it becomes imperative to consider not only the technical feasibility but also the broader implications of such advancements. Understanding how these enhancements might alter the very fabric of warfare requires a comprehensive examination of both the benefits and the risks involved.

Chapter 7.

Tactical Advantages and Battlefield Realities

In the following pages I would like you to consider two battle scenarios. The first battle takes place between two powerful nations. Both nations have a large military presence but one has a very significant strategic advantage over the other. This nation has implemented the use of genetically modified CRISPR soldiers.

As dawn breaks over the contested territory, the traditional forces of both nations are deployed, with tanks and infantry lining the borders. However, the nation with CRISPR-enhanced soldiers, referred to as Nation A, has a hidden ace up its sleeve. In the dense forests and rugged mountains, their genetically modified troops, known as the Vanguard, prepare for an unprecedented offensive.

The Vanguard soldiers possess enhancements that push human capabilities to their limits. Their muscle fibers are denser, allowing them to carry heavier loads and move with greater speed and agility. Enhanced vision and hearing give them superior situational awareness, and their modified cognitive functions enable faster decision-making and improved strategic planning. Their resistance to extreme temperatures and radiation makes them almost invincible in harsh environments.

As the first light of dawn spills over the horizon, the air is thick with tension. The rumble of tanks and the march of boots create a symphony of impending conflict. The traditional troops of both nations stand ready, their faces a mix of determination and apprehension. Across the barren no-man's land, the opposing forces of Nation A and Nation B are poised for battle, their lines stretching as far as the eye can see.

The silence is shattered by the deafening roar of artillery fire. Explosions rip through the earth, sending plumes of smoke and debris skyward. The conventional soldiers of both armies charge forward, their uniforms blending into a chaotic sea of movement. Gunfire erupts, the staccato bursts punctuating the cries of the wounded and the clash of metal on metal.

Amidst this traditional chaos, a new and unsettling presence makes itself known. Emerging from the shadows of the forest and the rugged terrain are the Vanguard soldiers of Nation A. Clad in sleek, high-tech combat suits, they move with an unnatural fluidity, their every motion precise and calculated. These are not ordinary soldiers; they are the pinnacle of genetic engineering, designed for perfection in warfare.

The Vanguard soldiers advance, their neural interfaces allowing them to communicate and coordinate with a level of efficiency that seems almost telepathic. Without a word, they split into squads, each unit moving towards a specific target with unwavering focus. Their enhanced vision cuts through the smoke and dust, identifying enemy positions with pinpoint accuracy.

As they engage the enemy, the difference becomes immediately apparent. The Vanguard's enhanced strength allows them to carry heavy weaponry effortlessly, their movements swift and powerful. They leap over obstacles with ease, their agility unmatched by any conventional soldier. In hand-to-hand combat, they are unstoppable, their blows shattering bones and disabling opponents in seconds.

In the midst of the chaos, another stark difference becomes glaringly obvious: the Vanguard soldiers' perfect accuracy when firing their weapons. Each shot is calculated and precise, their neural interfaces guiding their aim with unparalleled precision. Unlike the traditional soldiers, who often fire recklessly in the heat of battle, the Vanguard's genetically enhanced vision and steady hands ensure that every bullet finds its mark. The death toll among Nation B's troops skyrockets as the Vanguard pick off their targets with clinical efficiency, each shot a fatal blow that minimizes waste and maximizes lethality. Ammunition, once expended in torrents of suppressive fire, is now used sparingly and with deadly purpose. The battlefield is eerily devoid of the cacophony of continuous gunfire, replaced instead by the measured, devastatingly effective bursts of the Vanguard's weapons. This new efficiency in warfare not only leads to higher death rates among the enemy but also conserves resources, showcasing the brutal efficacy of CRISPR-enhanced soldiers and their transformative impact on the art of war.

The enemy lines, once solid and formidable, begin to crumble under the onslaught. Vanguard soldiers breach defenses with relentless force, their neural interfaces enabling perfect synchronization in their attacks. One squad moves through a trench, neutralizing every enemy soldier with ruthless efficiency. Another team takes out a fortified position, their coordinated strikes leaving no room for retaliation.

From a distance, commanders of Nation B watch in horror as their forces falter. The Vanguard's superior speed allows them to outmaneuver their enemies, flanking positions and cutting off escape routes. Their resilience is astounding; wounds that would incapacitate a normal soldier barely slow them down, their bodies rapidly healing as they continue their assault. In one harrowing instance, a Vanguard soldier takes a direct hit to the chest from an enemy sniper. The force of the impact staggers him momentarily, but within seconds, the genetically enhanced tissues begin to mend. The soldier straightens, his expression unchanged, and resumes his relentless advance. The sight of a seemingly invincible adversary recovering so swiftly from a potentially fatal injury further demoralizes Nation B's troops, reinforcing the terrifying power of Nation A's CRISPR-enhanced warriors.

The psychological impact on Nation B's soldiers is devastating. The sight of the Vanguard – these seemingly invincible warriors – shatters their morale. Fear and confusion spread like wildfire through the ranks. Soldiers who once stood firm now break and run, unable to comprehend the might of their genetically enhanced adversaries. The battlefield, already a scene of chaos, descends further into disarray.

In the midst of this carnage, the Vanguard soldiers maintain an eerie calm, their movements precise and unhurried despite the chaos around them. Their lack of empathy, a side effect of their genetic conditioning, renders them efficient but merciless combatants. Eyes devoid of emotion, they press forward unflinching, their only objective to annihilate the enemy. The battlefield is a symphony of violence, yet the Vanguard moves with a disconcerting serenity, as if they are merely executing a well-rehearsed drill.

One soldier, his face expressionless, dispatches enemies with surgical precision. His enhanced vision allows him to see through the smoke and debris, picking off targets with lethal accuracy. Another Vanguard, a woman with augmented strength, tears through the enemy lines, her blows breaking bones and ending lives with mechanical efficiency.

Civilians caught in the crossfire receive no mercy. An elderly woman, clutching a child, stumbles into the open, her face a mask of terror. A Vanguard soldier, his neural interface prioritizing threats, raises his weapon without a moment's hesitation. A single, silenced shot ends their lives, the soldier moving on without a backward glance. The Vanguard's programming is devoid of compassion or hesitation, their actions dictated by cold logic and unyielding orders.

Their calm demeanor contrasts sharply with the panic and desperation of Nation B's soldiers. Where the Vanguard is methodical, the traditional troops are frantic, their lines crumbling under the relentless assault. The sight of their comrades falling with horrifying efficiency breaks their spirits, the psychological impact of facing such implacable foes proving too much to bear.

As the Vanguard presses forward, their lack of hesitation and unfeeling precision create an atmosphere of dread. They are not just soldiers; they are instruments of destruction, crafted with the sole purpose of ensuring victory at any cost. The battlefield, once a place of human struggle and bravery, has become a stage for a new kind of warfare—one where humanity itself is a liability, and the future belongs to those who can transcend its limitations.

The soldiers of the opposing nation, Nation B, fight valiantly but are no match for the genetic super soldiers. The Vanguard can operate for extended periods without rest, thanks to their enhanced endurance and rapid healing capabilities. Even when injured, their bodies regenerate quickly, allowing them to return to the fight with minimal downtime. Nation B's soldiers, witnessing the apparent invincibility of their adversaries, begin to lose morale, the psychological impact of facing such formidable opponents taking its toll.

As the battle rages on, the true extent of the genetic modifications becomes evident. The Vanguard soldiers demonstrate not only physical prowess but also a chilling detachment. Their lack of empathy, a side effect of their genetic conditioning, makes them ruthless in combat. They execute their orders without hesitation, targeting enemy combatants and civilians alike with a cold, mechanical efficiency. The battlefield becomes a grim tableau of human and ethical devastation, the line between soldier and machine blurring with each passing moment.

Despite their technological superiority, the CRISPR-enhanced soldiers face their own set of challenges. The psychological toll of their modifications, coupled with the neural interfaces' constant demands, begins to manifest in unexpected ways. Some soldiers experience mental breakdowns, their minds unable to cope with the relentless strain. Others become increasingly desensitized, their humanity eroding as the enhancements take over. The very modifications that make them super soldiers also isolate them from the rest of society, creating a new class of beings who are feared and misunderstood.

The implications of this battle reach far beyond the immediate conflict. Nation A's victory, achieved through the use of genetically modified soldiers, sets a dangerous precedent. Other nations, recognizing the strategic advantage, begin their own CRISPR programs, sparking a global arms race in genetic enhancement. The ethical debates that once dominated scientific discussions are drowned out by the clamor for military supremacy. The world stands on the brink of a new era, where the boundaries of human potential are pushed to their extremes, and the consequences of playing god with genetics become terrifyingly real.

In this dystopian future, the pursuit of genetic superiority has redefined the nature of warfare and humanity itself. The battle between Nation A and Nation B is just the beginning, a harbinger of the profound changes that lie ahead. As nations vie for dominance in this new frontier, the price of victory becomes increasingly steep, and the line between human and machine continues to blur, leaving society to grapple with the ethical and existential ramifications of a world forever altered by genetic engineering.

Now that you have finished reading the first scenario, consider the second one. Another war has erupted between two major countries. This time, however, both nations have deployed soldiers genetically enhanced using CRISPR technology.

As the sun rises over the battlefield, the air is heavy with anticipation. Two powerful nations, Nation X and Nation Y, stand on the brink of war. Unlike the previous conflict, both sides have deployed their own genetically enhanced soldiers, each boasting the formidable capabilities granted by CRISPR technology. The world watches with bated breath, aware that this clash between super-soldiers could reshape the future of warfare—and humanity itself.

The initial skirmishes are brutal and swift. Nation X's elite forces, known as the Titans, move with an almost supernatural speed and agility. Their genetically modified muscles propel them through the terrain with ease, while their enhanced reflexes make them nearly impossible to hit. On the other side, Nation Y's soldiers, the Phantoms, are equally formidable. Their augmented cognitive functions allow them to anticipate enemy movements and react with lightning-fast precision.

The battlefield becomes a theater of high-tech carnage. Titans and Phantoms clash in a series of brutal engagements, their superhuman abilities pushing both sides to the brink. The Titans' strength and resilience are countered by the Phantoms' superior strategy and enhanced sensory perception. Each side employs advanced tactics, using their genetic modifications to gain the upper hand. Yet, as the battle rages on, it becomes clear that neither side can achieve a decisive victory.

The enhanced soldiers, while extraordinarily powerful, also embody the weaknesses of their enhancements. The Titans' relentless aggression makes them prone to overextending themselves, while the Phantoms' reliance on strategy sometimes delays their immediate response to unexpected threats. Both nations suffer heavy casualties, their super-soldiers falling in droves despite their genetic advantages.

The scale of destruction escalates rapidly. Entire cities are leveled as the combat spills over into civilian areas. Infrastructure collapses under the relentless assault, and the death toll rises with horrifying speed. The genetic enhancements that were meant to provide an edge become instruments of mutual annihilation. The battlefield is littered with the bodies of fallen Titans and Phantoms, a grim testament to the devastating power of human ingenuity turned destructive.

As the conflict drags on, the initial optimism surrounding the use of CRISPR-enhanced soldiers fades. The costs are too high, the destruction too widespread. Both nations begin to realize that their pursuit of genetic superiority has led them down a path of no return. The environmental impact is catastrophic; radiation leaks from destroyed power plants, and toxic chemicals from devastated industrial areas contaminate the land and water. The very fabric of society starts to unravel under the weight of continuous warfare.

The leaders of Nation X and Nation Y, once confident in their technological prowess, now face the grim reality of their decisions. In a desperate bid to end the bloodshed, they resort to more drastic measures. Cyberattacks and biological weapons are deployed, further accelerating the spiral into mutual destruction. The enhanced soldiers, now mere pawns in a game of global annihilation, continue to fight, their human instincts for survival overridden by their genetic programming.

In the end, the war leaves both nations in ruins. The once-thriving cities are reduced to rubble, and the population is decimated. The few survivors, both military and civilian, are left to grapple with the aftermath of a conflict that has irreparably damaged their world. The dream of creating super-soldiers to secure victory has instead wrought a nightmare of mutual destruction, a stark reminder of the perils of unchecked technological advancement.

As the dust settles, the world looks on in horror, recognizing that the pursuit of genetic enhancement has not led to superiority but to desolation. The story of the Titans and the Phantoms serves as a chilling warning of the consequences of playing god with human genetics, a dark chapter in human history that underscores the need for restraint, ethical consideration, and a renewed commitment to peace.

In considering the implications of deploying CRISPR-enhanced soldiers on the battlefield, it's essential to envision how these hypothetical conflict scenarios might unfold. The potential for these genetically modified troops to reshape military engagements is immense, driven by the combination of advanced genetic engineering and traditional combat strategies.

Imagine a near-future conflict between two technologically advanced nations. Both sides have developed CRISPR-modified soldiers, designed for superior physical and cognitive abilities. These soldiers are not just stronger and faster; they possess enhanced night vision, resistance to fatigue, and accelerated healing capabilities. In such a scenario, traditional concepts of warfare would be radically altered.

During a high-stakes operation, a platoon of CRISPR soldiers could be deployed to infiltrate an enemy base located in a hostile, remote area. Their enhanced endurance would allow them to traverse harsh terrains without the need for extended rest. Upon reaching their target, these soldiers would utilize their augmented reflexes and coordination to execute precise, synchronized attacks, neutralizing threats with efficiency that far surpasses ordinary human capabilities.

The strategic advantages of such soldiers are not limited to physical enhancements. Cognitive improvements, such as heightened situational awareness and problem-solving skills, would enable CRISPR soldiers to adapt to rapidly changing battlefield conditions. For instance, if an unexpected enemy force arrives, these soldiers could quickly devise and execute complex tactical maneuvers, ensuring minimal casualties and mission success.

One of the most significant implications of deploying CRISPR soldiers lies in their potential to reduce the unpredictability of human factors in warfare. Traditional soldiers are subject to stress, fear, and fatigue, all of which can impair judgment and performance. In contrast, genetically enhanced soldiers could be engineered to withstand extreme psychological pressures, maintaining composure and effectiveness even in the most dire situations.

The psychological impact on opposing forces must also be considered. The presence of seemingly invincible soldiers could demoralize enemy troops, potentially leading to quicker surrenders or disorganized retreats. Historical parallels can be drawn to the use of elite units in past conflicts, where the fear and awe inspired by highly trained soldiers had a profound impact on the enemy's morale.

However, the deployment of CRISPR soldiers raises critical ethical and strategic questions. For example, what happens if these enhanced soldiers are captured by the enemy? The potential consequences are alarming. The capture of genetically modified soldiers could lead to the extraction and reverse engineering of their genetic modifications, allowing adversaries to replicate or even improve upon these enhancements. This scenario poses a significant risk of genetic information falling into the hands of hostile entities, who could then develop their own super-soldiers or create biological weapons designed to target specific genetic traits.

Such a development could spur a new kind of arms race, not based on nuclear or conventional weapons, but on genetic warfare. Nations would be compelled to invest heavily in genetic research and development to keep pace with their adversaries, leading to a rapid escalation of conflicts. The focus of military superiority would shift from traditional means to advanced biotechnological capabilities, with countries racing to outdo each other in creating the most enhanced and formidable soldiers.

The ethical and strategic implications of this arms race are profound. The secrecy and speed at which genetic technologies can be developed could lead to a lack of transparency and oversight, increasing the likelihood of accidental or intentional misuse. Moreover, the existence of genetically enhanced soldiers on the battlefield could destabilize global security dynamics, as nations without such capabilities might feel threatened and resort to unconventional or asymmetric warfare tactics.

In addition to the direct military consequences, there is also the risk of a broader societal impact. The proliferation of genetic enhancement technologies could lead to a black market for genetic modifications, where non-state actors, including terrorist groups, could gain access to these powerful tools. This would not only make it harder to control the spread of such technologies but also increase the potential for their use in acts of terrorism, further exacerbating global instability.

Furthermore, the ethical dilemmas surrounding the capture and exploitation of genetically modified soldiers would challenge existing norms of international humanitarian law. The treatment of these soldiers as mere sources of genetic data, rather than as human beings with rights and dignity, could lead to significant human rights violations. International bodies would need to develop new frameworks and regulations to address these issues, ensuring that the use of genetic technologies in warfare does not undermine the principles of human rights and ethical conduct.

The capture of enhanced soldiers by the enemy presents a cascade of potential risks and challenges that extend far beyond the battlefield. It highlights the need for careful consideration and regulation of genetic technologies in the military domain to prevent an uncontrolled escalation of genetic warfare and to protect global security and ethical standards.

While the benefits of CRISPR soldiers in combat are clear, the broader implications for warfare and international stability are complex. The ability to engineer super-soldiers could lead to a new era of conflict, where the lines between human and machine blur, and the ethical boundaries of warfare are continually tested. As we explore these hypothetical scenarios, it becomes evident that the advent of CRISPR technology in the military domain is not just a technological revolution, but a profound shift in the nature of human conflict.

Beyond the battlefield enhancements, genetic modifications could also address some of the most persistent challenges faced by military personnel, such as the psychological toll of warfare. Post-Traumatic Stress Disorder (PTSD) and other stress-related conditions are highly prevalent among soldiers, often resulting in long-term mental health issues and impairing operational effectiveness. By targeting genes that influence stress responses, such as those involved in the hypothalamic-pituitary-adrenal (HPA) axis, CRISPR technology could help soldiers manage stress more effectively. This could maintain mental health and operational readiness over extended periods of deployment, offering a holistic approach to enhancing military performance and well-being.

The HPA axis plays a crucial role in the body's response to stress, regulating the production of cortisol and other stress hormones. By using CRISPR to edit genes associated with the HPA axis, it may be possible to enhance a soldier's ability to cope with high-stress environments. For example, modifying the genes that regulate cortisol production could help maintain optimal hormone levels, preventing the detrimental effects of chronic stress, such as anxiety, depression, and impaired cognitive function.

Beyond cortisol regulation, genetic modifications could also target other pathways and molecules involved in stress responses. For instance, enhancing the expression of genes that promote the production of neuropeptides like oxytocin and vasopressin could bolster social bonding and resilience, providing soldiers with stronger support networks and improving their ability to handle the psychological demands of combat.

Furthermore, genetic interventions could be designed to increase neuroplasticity, the brain's ability to adapt and reorganize itself in response to new experiences and challenges. By promoting neuroplasticity, CRISPR could help soldiers recover more quickly from traumatic events, reducing the incidence and severity of PTSD. This approach could involve editing genes that influence the production of brain-derived neurotrophic factor (BDNF), a protein that supports neuron growth and connectivity.

Another promising avenue for reducing the psychological toll of warfare through genetic modification involves enhancing the brain's natural reward systems. By targeting genes that regulate the dopamine and serotonin pathways, it may be possible to improve mood regulation and motivation, helping soldiers maintain a positive outlook even in adverse conditions. This could not only enhance their mental resilience but also improve overall morale and unit cohesion.

The potential benefits of such genetic modifications extend beyond individual soldiers, potentially leading to a more resilient and effective military force. By reducing the prevalence of PTSD and other stress-related conditions, these interventions could lower the long-term healthcare costs associated with treating veterans, as well as improve retention rates by reducing the number of personnel who leave the military due to mental health issues.

While the ethical implications of genetically modifying soldiers must be carefully considered, the potential to significantly improve their mental health and operational readiness presents a compelling argument for further research and development in this area. As the technology advances, it is crucial to establish stringent ethical guidelines and oversight mechanisms to ensure that these genetic modifications are used responsibly and with the best interests of the soldiers in mind.

Moreover, CRISPR soldiers themselves may face unique psychological challenges. The process of genetic enhancement, while providing physical benefits, can lead to identity crises and ethical dilemmas. Soldiers might struggle with questions about their humanity and individuality, knowing they have been engineered for war. Studies on soldiers who have undergone extensive physical and psychological training, such as those in special forces, indicate that extreme conditioning can lead to identity conflicts and mental health issues. It stands to reason that genetic enhancements could amplify these effects, creating a class of warriors who are physically superior but mentally burdened by their enhancements. Furthermore, soldiers enhanced by CRISPR might experience isolation from their peers and society due to their perceived differences, exacerbating feelings of alienation and depression.

These psychological and social challenges are not just individual issues but have broader implications for military operations and strategy. Tactically, the introduction of CRISPR soldiers could redefine battlefield strategies. Traditional tactics may become obsolete when facing enemies who can recover from injuries faster, carry heavier loads, and endure extreme conditions.

Military strategists will need to develop new doctrines that leverage the capabilities of enhanced soldiers while countering those of the enemy. For instance, urban warfare, which relies heavily on the physical endurance and adaptability of soldiers, could see a shift in tactics to account for the heightened abilities of CRISPR troops. Enhanced soldiers could perform prolonged operations without the need for rest, drastically altering the timing and tempo of engagements. This could force adversaries to rethink their approach to warfare, possibly leading to a new era of conflict characterized by rapid, relentless assaults. Thus, the psychological burdens faced by CRISPR soldiers are intertwined with the strategic adjustments required on the battlefield, highlighting the complex impact of genetic enhancements on modern warfare.

The logistical aspects of warfare would also be profoundly impacted by the introduction of genetically enhanced soldiers. Traditional supply lines, which currently account for food, medical supplies, and other necessities, would need to be fundamentally reconsidered. Enhanced soldiers, with their altered dietary needs and superior healing capabilities, could reduce the dependency on conventional rations and medical care. This shift has the potential to streamline operations by reducing the volume and variety of supplies that need to be transported to and maintained at the front lines.

For example, genetically modified soldiers might require specially formulated nutrient packs designed to sustain their enhanced metabolic processes and support their advanced physical capabilities. These nutrient packs could replace traditional food supplies, significantly lightening the logistical load. Imagine soldiers who have undergone genetic modifications to boost muscle mass and endurance through enhanced expressions of genes like MSTN (myostatin) inhibitors or PGC-1alpha (a regulator of mitochondrial biogenesis). Such modifications would result in increased metabolic rates and higher nutritional demands to support their advanced physical states.

The nutrient packs for these soldiers would need to contain a precise balance of macronutrients, vitamins, and minerals tailored to their specific needs. For instance, these packs could be rich in high-quality proteins to aid muscle repair and growth, complex carbohydrates for sustained energy, and essential fatty acids to support cognitive functions. Additionally, they might include increased amounts of electrolytes to prevent dehydration and enhance muscle function, as well as specific amino acids like leucine, which is critical for muscle protein synthesis.

These nutrient packs could be further enhanced with bioengineered compounds designed to maximize absorption and utilization. For instance, they might include nanotechnology-based delivery systems that ensure nutrients are released in a controlled manner, providing sustained energy and nutrient supply throughout extended missions. This technology could help maintain optimal physical and cognitive performance even under the most demanding conditions.

Another hypothetical example involves genetically modified soldiers who possess enhanced healing capabilities, achieved by modifying genes like VEGF (vascular endothelial growth factor) to promote rapid tissue repair and regeneration. These soldiers would benefit from nutrient packs that contain ingredients specifically formulated to support accelerated healing processes. Such packs could include high concentrations of vitamins A and C, which are essential for collagen synthesis and wound healing, as well as zinc and magnesium, which play critical roles in tissue repair.

Moreover, these nutrient packs could be designed to provide anti-inflammatory compounds, such as omega-3 fatty acids and curcumin, to reduce recovery time and minimize downtime due to injuries. The inclusion of probiotics and prebiotics might also be considered to support gut health and enhance the immune system, ensuring that soldiers remain healthy and resilient in various environments.

By replacing traditional food supplies with these advanced nutrient packs, the logistical burden of transporting and storing perishable food items could be significantly reduced. This would allow for more efficient use of resources and greater flexibility in planning and executing military operations. The nutrient packs could be lightweight, compact, and have a long shelf life, making them ideal for deployment in remote or hostile environments where resupplying might be challenging.

In addition to enhancing physical capabilities, these nutrient packs could be designed to support cognitive functions in genetically modified soldiers. For example, soldiers with enhanced cognitive abilities through modifications in genes like BDNF (brain-derived neurotrophic factor) or COMT (catechol-O-methyltransferase) would require nutrients that support brain health and function. These packs could include omega-3 fatty acids, known for their neuroprotective properties, as well as antioxidants like resveratrol and flavonoids to protect against oxidative stress.

Furthermore, the inclusion of nootropic compounds, such as phosphatidylserine and acetyl-L-carnitine, could enhance memory, focus, and cognitive endurance. Adaptogens like rhodiola and ashwagandha might also be added to help soldiers manage stress and maintain mental clarity during prolonged missions.

The development of specially formulated nutrient packs for genetically modified soldiers would not only support their advanced physical and cognitive capabilities but also streamline logistical operations. By providing targeted nutrition that meets the unique needs of these enhanced soldiers, military forces could ensure that they remain at peak performance, ready to face the challenges of modern warfare.

However, this logistical streamlining introduces new vulnerabilities. The dependence on specialized nutrients and genetic maintenance protocols means that any disruption in the supply chain could have catastrophic effects on the combat readiness of these enhanced troops. If the delivery of specific nutrients essential to their modified biology were interrupted, it could lead to malnutrition and a rapid decline in their physical capabilities. Similarly, if the mechanisms required to maintain their genetic modifications—such as regular doses of gene-editing enzymes or suppressors—were compromised, the enhancements could fail, leaving soldiers not only ineffective but possibly suffering from severe physiological repercussions.

Furthermore, the logistics of maintaining the integrity of genetic modifications in the field would present unique challenges. Equipment and personnel trained to administer and monitor these modifications would be crucial. This introduces another layer of complexity to military operations, as ensuring the security and functionality of this specialized support system becomes as critical as maintaining traditional supply lines.

The potential for targeted attacks on these new supply chains could become a significant strategic concern. Adversaries might focus efforts on disrupting the specialized logistical operations to cripple the enhanced forces, knowing that the failure of these systems could lead to a rapid decline in the effectiveness of genetically modified soldiers.

Thus, while the logistical implications of deploying genetically enhanced soldiers could offer streamlined operations and potentially significant advantages, they also require robust planning and safeguards to prevent new types of vulnerabilities from being exploited. The balance between the benefits of reduced traditional logistical needs and the risks associated with specialized requirements will be a critical factor in the successful integration of genetically enhanced soldiers into modern military strategies.

Resistance movements and insurgencies might also adapt their strategies in response to the new capabilities of CRISPR soldiers. Guerrilla tactics that rely on the element of surprise and the physical limitations of conventional troops might become less effective. Insurgents could instead focus on cyber warfare, targeting the technology and infrastructure supporting genetic enhancements. This could usher in a new era of hybrid warfare, where traditional combat is combined with sophisticated technological attacks.

As the dust settles on these hypothetical battlefields, it becomes clear that the psychological and tactical impacts of CRISPR soldiers are complex and far-reaching. While they offer significant advantages, they also bring about new challenges and ethical considerations. This exploration sets the stage for a deeper examination of how society perceives these genetically enhanced warriors and the cultural ramifications of their existence. The implications extend beyond the battlefield, influencing public opinion, policy-making, and the very fabric of society.

Chapter 8.

Public Perception and Media Influence

In recent years, media portrayal of technological advancements has had a profound impact on public perception. As we delve into the potential media representation of CRISPR soldiers, it's essential to consider both the sensationalism that often accompanies such coverage and the nuanced discussions that can shape public opinion.

Media outlets thrive on capturing attention, and the concept of genetically modified soldiers offers a wealth of sensational material. Headlines might scream about "super-soldiers" with extraordinary abilities, playing into both awe and fear. Such portrayals are not new; history shows that the media has often exaggerated scientific advancements. For example, during the early days of genetic engineering, the media frequently focused on the potential for "designer babies," creating a mix of excitement and ethical panic among the public. Headlines often proclaimed the possibility of parents selecting their child's traits, from eye color and intelligence to athletic prowess and disease resistance. This notion captured the imagination of many, promising a future where hereditary diseases could be eradicated and human potential optimized from birth.

Alongside the excitement came significant ethical concerns. The idea of "designer babies" raised profound questions about the moral implications of genetic manipulation. Critics argued that such capabilities could lead to a new form of eugenics, where societal pressures might dictate what constitutes a "desirable" trait, potentially exacerbating social inequalities and discrimination. They feared a future where only the wealthy could afford genetic enhancements for their children, creating a genetic underclass and deepening the divide between the haves and the have-nots.

Moreover, the potential for unforeseen consequences loomed large. The genetic editing of embryos posed risks of unintended mutations and long-term effects that could not be predicted or controlled. This uncertainty fueled a broader debate about the extent to which humanity should interfere with natural genetic processes and the potential hubris of "playing God."

As a result, the public discourse became a battleground of futuristic optimism and ethical trepidation. While some envisioned a new era of human health and capability, others warned of the slippery slope towards a dystopian future. The intense media focus on "designer babies" thus played a critical role in shaping public perception and policy discussions, highlighting the need for robust ethical frameworks and regulations to guide the responsible use of genetic engineering technologies.

Television shows, movies, and books have long explored the idea of enhanced humans, from comic book superheroes to dystopian futures. These portrayals tend to oscillate between glorification and dystopian warnings. For instance, films like "Gattaca" and "Captain America: The First Avenger" illustrate both the allure and the ethical quandaries of genetic enhancement. These cultural narratives influence how society might view real-life CRISPR soldiers, likely casting them as either heroic saviors or dangerous aberrations.

Discussions in the media about genetically modified soldiers are not yet commonplace, as the topic is still emerging and largely speculative. However, several notable examples in mainstream media touch on the broader themes of genetic modification and military applications, including potential enhancements for soldiers.

Vice produced an intriguing documentary titled "Super Soldiers" that delves into the future of warfare and the potential for creating enhanced soldiers through advanced technologies, including genetic engineering. The documentary offers a comprehensive exploration of this futuristic concept by featuring in-depth interviews with military experts, bioethicists, and scientists. These interviews shed light on the technological possibilities and ethical implications of creating super soldiers, providing viewers with a balanced perspective on this controversial topic.

"Super Soldiers" begins by examining the historical context of human enhancement in the military, tracing the evolution from performance-enhancing drugs and rigorous training programs to the current frontier of genetic engineering. The documentary highlights the significant investments made by the Pentagon and DARPA (Defense Advanced Research Projects Agency) in research aimed at enhancing human capabilities. It specifically references DARPA's Biological Technologies Office and its substantial budget dedicated to exploring the intersection of biology and engineering.

Through interviews with bioethicists, "Super Soldiers" addresses the ethical concerns surrounding the enhancement of human beings for military purposes. These experts discuss the potential for misuse, the blurring of lines between therapeutic and enhancement applications, and the long-term societal impacts of creating genetically modified soldiers. The documentary raises critical questions about the moral and ethical boundaries of such technologies, emphasizing the need for stringent oversight and international regulations.

Moreover, the documentary features insights from scientists involved in genetic research, including those working on projects like the Human Genome Project and genome writing initiatives funded by DARPA. These projects aim to engineer human cells capable of manufacturing essential nutrients and self-sustaining in extreme environments, which could lead to the development of soldiers with minimal need for external sustenance.

"Super Soldiers" provides a thought-provoking and detailed exploration of the future of military technology, blending scientific possibilities with ethical considerations. It underscores the potential and challenges of genetic engineering in creating enhanced soldiers, prompting viewers to reflect on the profound implications of these advancements for the future of warfare and humanity itself.

National Geographic's "Breakthrough: More Than Human," an episode directed by Paul Giamatti, delves deeply into the realm of human enhancement through biotechnology, including the controversial and rapidly advancing field of genetic modifications. This compelling episode explores the potential for creating enhanced soldiers, a concept that has moved from the realm of science fiction into the forefront of military research and development.

The episode provides a comprehensive look at how genetic engineering, particularly CRISPR technology, is being explored to develop super-soldiers with superior physical and cognitive abilities. It features interviews with leading scientists, such as those involved in DARPA's cutting-edge projects, who discuss the technical advancements and possibilities that genetic modifications present. These experts explain how CRISPR can be used to edit genes responsible for muscle growth, endurance, and resistance to diseases and environmental stresses, aiming to create soldiers who are not only stronger and faster but also more resilient in extreme conditions.

Ethicists also play a significant role in the episode, providing a balanced view of the moral considerations surrounding these advancements. They raise critical questions about the implications of such technologies on human rights and the potential for misuse. The discussions highlight the thin line between therapeutic applications of genetic engineering and its use for enhancement, a distinction that becomes increasingly blurred as the technology advances.

The episode also touches on historical precedents, such as the Edgewood Arsenal experiments, where the U.S. military tested various chemical agents on soldiers, pushing ethical boundaries in the pursuit of enhanced capabilities. Conducted between 1948 and 1975, these experiments involved exposing military personnel to a range of chemical substances, including nerve agents, hallucinogens, and stimulants, with the goal of understanding their effects on human performance and developing potential chemical weapons. The participants, often not fully informed about the risks, experienced severe physical and psychological side effects, leading to significant ethical and legal ramifications.

This historical context underscores the importance of ethical oversight in contemporary genetic research, drawing clear parallels between past and present military enhancement efforts. Just as the

Edgewood Arsenal experiments pushed the limits of ethical conduct under the guise of national security and scientific advancement, current explorations into genetic modifications for soldiers, such as those involving CRISPR technology, tread similarly precarious ethical ground. These past abuses highlight the potential for misuse when scientific ambition is unchecked by moral considerations.

In the Edgewood Arsenal experiments, the lack of informed consent and the willingness to use soldiers as test subjects for high-risk research set a troubling precedent. These actions sparked public outrage and led to stricter regulations on human experimentation, exemplified by the establishment of the Belmont Report in 1979, which set ethical guidelines for research involving human subjects. The lessons from this era are stark reminders of the need for transparency, consent, and stringent ethical standards in any form of human experimentation.

Today's advancements in genetic engineering, particularly within military contexts, evoke these historical parallels. The pursuit of creating enhanced soldiers through CRISPR technology, which promises to improve physical strength, cognitive abilities, and resilience to environmental stressors, mirrors the same drive for superiority seen in past military research. However, the ethical implications of editing human genes are profound, raising questions about consent, long-term effects, and the potential for creating inequalities or unintended consequences.

The BBC has extensively covered DARPA's various projects aimed at enhancing soldier capabilities, providing detailed reports and analyses on the implications of these advancements. Notable among these is DARPA's Safe Genes program, which seeks to develop tools for controlling gene editing to prevent accidental or intentional misuse of genome editing technologies. The program aims to protect military personnel from unintended genetic changes and to facilitate the development of safe, precise, and effective medical treatments using gene editors. This initiative highlights DARPA's commitment to ensuring that the powerful capabilities of CRISPR technology are harnessed responsibly and ethically.

Similarly, PBS's acclaimed science series Nova produced a documentary titled "Genetically Modified Humans," which delves into the cutting-edge advancements in genetic engineering, with a significant focus on CRISPR technology. While the documentary primarily addresses the potential medical applications of genetic modifications, it also explores the broader implications, including the possibility of using such technologies for military purposes. The documentary raises critical questions about the ethical and moral boundaries of using CRISPR to enhance human abilities, particularly in the context of creating super soldiers.

As media coverage continues to highlight these advancements, the ethical debates surrounding the use of genetic technology for military purposes grow increasingly complex. The

BBC, PBS, and other outlets play a crucial role in informing the public and fostering dialogue on the potential benefits and risks of these technologies. By examining both the scientific potential and the ethical challenges, these reports and documentaries contribute to a broader understanding of the implications of using CRISPR and other genetic technologies in the pursuit of enhanced human capabilities.

The Guardian has published several in-depth articles exploring the intersection of bioengineering and military applications, frequently highlighting expert opinions on the ethical and societal implications of using genetic modification to create enhanced soldiers. These articles provide a balanced view, weighing the potential benefits against the risks and moral concerns associated with genetic enhancements in military contexts. For instance, The Guardian has featured insights from bioethicists, military analysts, and genetic researchers who discuss the profound consequences of deploying genetically modified soldiers. The discussions often revolve around the potential for increased military effectiveness and the significant ethical dilemmas, such as the loss of individual autonomy and the possibility of unintended genetic consequences.

Scientific American has also extensively covered the topic of genetic enhancement, particularly focusing on its potential applications within the military. Articles in Scientific American delve into the scientific advancements in CRISPR and other gene-editing technologies, offering detailed explanations of how these innovations can be harnessed to enhance human performance. The publication explores the practical considerations of implementing such technologies in military settings, including the logistical challenges and the need for rigorous testing to ensure safety and efficacy. Additionally, Scientific American addresses the ethical considerations, emphasizing the necessity of developing robust regulatory frameworks to prevent misuse and to ensure that genetic modifications are conducted ethically and responsibly.

One notable discussion in Scientific American includes the potential for CRISPR to enhance muscle growth, improve cognitive functions, and increase resistance to environmental stresses such as extreme temperatures and radiation. These enhancements could lead to the creation of soldiers who are not only physically superior but also capable of enduring and thriving in conditions that would typically be debilitating. The articles highlight ongoing research projects, such as those funded by DARPA, which aim to explore these possibilities. DARPA's Safe Genes program, for example, seeks to develop tools to control gene editing, ensuring that any genetic modifications can be precisely targeted and reversed if necessary. This program reflects a broader commitment to balancing innovation with safety, aiming to protect military personnel from the potential risks associated with genome editing.

Furthermore, both The Guardian and Scientific American discuss the broader societal implications of genetic enhancements.

They consider scenarios in which the line between therapeutic use and enhancement becomes blurred, raising concerns about equity and access to such technologies. The possibility of creating a genetically enhanced class of soldiers prompts questions about the long-term impact on society and the potential for new forms of inequality and discrimination.

In summary, both publications provide comprehensive coverage of the advancements in genetic engineering and their potential military applications, offering a nuanced view of the benefits and challenges. Their articles underscore the need for ongoing dialogue and ethical scrutiny as these technologies continue to evolve, ensuring that the pursuit of enhanced human capabilities does not come at the cost of ethical integrity and societal well-being.

These examples highlight how mainstream media outlets are beginning to explore the complex and controversial topic of genetically modified soldiers. Through documentaries, news reports, and articles, these media pieces provide a platform for experts to discuss the technological possibilities and ethical dilemmas associated with this emerging field. As research progresses and the implications of genetic modification become more apparent, it is likely that media coverage on this topic will continue to grow.

Ultimately, the portrayal of CRISPR soldiers in news reports and documentaries would play a crucial role in shaping public discourse. By presenting a diverse range of viewpoints and carefully framing the discussion, media can help foster a nuanced understanding of the complex ethical, societal, and scientific dimensions of this emerging technology. This balanced presentation is vital as it ensures that the public remains informed and engaged with the potential benefits and risks associated with genetic modifications.

In a related development reported by the South China Morning Post, a Chinese military research team has conducted a groundbreaking experiment by inserting a gene from the water bear (tardigrade) into human embryonic stem cells, significantly boosting their resistance to radiation. Using CRISPR/Cas9 technology, nearly 90% of the modified cells survived lethal X-ray exposure. This achievement raises the possibility of creating super soldiers who can endure extreme conditions, such as nuclear fallout. This experiment exemplifies the type of scientific advancements that necessitate thorough media coverage to explore their implications comprehensively.

The water bear is renowned for its extraordinary resilience, capable of surviving extreme temperatures, radiation, and even the vacuum of space. By integrating the water bear's damage-suppressor protein gene into human cells, the researchers aimed to transfer these survival traits to humans. The experiment has captured widespread media attention and sparked intense discussions within the scientific community and beyond.

While the scientific community acknowledges the potential benefits, such as enhanced survivability of soldiers in hostile environments, there are significant ethical and safety concerns. The long-term effects of such genetic modifications are unknown, and there is a risk of harmful mutations and unforeseen health issues. The cross-species gene transfer could have unpredictable biological consequences, making it imperative to conduct thorough and long-term studies to ensure safety.

The media coverage reflects a mix of intrigue and concern. Outlets like the South China Morning Post have highlighted the technical achievements and the potential military applications, while also noting the ethical quandaries and the possibility of misuse. Discussions on forums and news sites emphasize the unprecedented nature of this experiment and the potential for both positive applications and serious ethical dilemmas.

Public reaction has been varied. Some view the research as a remarkable scientific advancement with potential to protect soldiers and civilians in nuclear scenarios. Others are alarmed by the ethical implications, fearing the creation of a new arms race focused on genetically enhanced humans. The debate highlights the need for stringent ethical guidelines and international regulations to govern the use of such powerful biotechnologies.

Social media will also play a crucial role in shaping public perception. Platforms like Twitter, Facebook, and YouTube enable rapid dissemination of information and opinions. Viral posts, memes, and videos can amplify both accurate information and misinformation. For example, the rapid spread of conspiracy theories during the COVID-19 pandemic demonstrates how quickly narratives can take hold, regardless of their basis in fact. Hashtags and trending topics related to CRISPR soldiers would likely follow a similar pattern, with polarized views emerging and influencing the broader public discourse.

It's essential to acknowledge the potential for misinformation and sensationalism to overshadow nuanced discussions. The media's focus on dramatic and extreme scenarios could lead to an overestimation of the immediate capabilities and risks of CRISPR technology. Balanced and responsible journalism will be crucial in ensuring that the public receives accurate information, fostering informed and rational discussions about the ethical and practical implications of CRISPR soldiers.

The media's portrayal of CRISPR soldiers will be a complex interplay of sensationalism, ethical debates, and cultural narratives. As we move forward, it will be vital to critically engage with these portrayals and seek out balanced perspectives to navigate the intricate landscape of genetic advancements and their impact on society.

Chapter 9.

Cultural and Societal Impacts

In the evolving discourse around CRISPR technology and its applications in warfare, cultural reactions play a crucial role. Societies around the globe have distinct perspectives shaped by historical, ethical, and sociopolitical contexts. These reactions are not monolithic; they are a tapestry of diverse viewpoints reflecting deep-seated values and concerns.

In many Western nations, particularly those with a history of liberal democracy, the reaction to genetic modification in warfare is often cautious and laden with ethical scrutiny. The public discourse frequently revolves around the moral implications of "playing God" and the potential for creating a new class of enhanced humans, reminiscent of dystopian fiction. Historical precedents, such as the eugenics movements of the early 20th century, have left a lingering suspicion towards genetic manipulation, particularly when linked to state power and military objectives. Public opinion in these regions is generally split, with a significant portion expressing concern over the potential for abuse and the erosion of individual rights.

In contrast, countries with strong centralized governments and a strategic focus on technological advancement, such as China, display a more pragmatic approach. The Chinese government has heavily invested in genetic research, viewing CRISPR technology as a critical component of their national defense strategy. This stance is often reflected in the broader societal acceptance of genetic modification, seen as a necessary evolution to maintain global competitiveness. However, this does not eliminate internal debates. Chinese intellectuals and scientists frequently engage in discussions about the ethical boundaries of such technologies, although these debates are often less visible to the global community due to governmental control over media and public discourse.

In regions like the Middle East, cultural and religious perspectives significantly influence the reception of genetic modification in warfare. Islamic bioethics, which emphasize the sanctity of the human body as created by God, generally exhibit strong resistance to genetic modifications. These views are deeply ingrained in the societal fabric and inform both public opinion and policy-making. For instance, in countries like Saudi Arabia and Iran, religious leaders wield considerable influence and often voice strong opposition to genetic engineering, framing it as an affront to divine will. This creates a complex landscape where technological advancement must navigate the stringent ethical boundaries set by religious doctrines.

African nations present a varied picture influenced by their unique historical and socio-economic contexts. In many African countries, the immediate concerns are more about access to basic healthcare and technology rather than the high-tech advancements represented by CRISPR. However, there is a growing awareness and curiosity about genetic technology, especially among the younger, more educated population. This demographic sees potential benefits in areas like disease eradication and agricultural improvement, but the militarization of genetic technology is often viewed with suspicion. Memories of colonial exploitation and unethical medical practices, such as the infamous Tuskegee Syphilis Study, contribute to a wary approach towards any form of genetic experimentation.

These cultural reactions are not static; they evolve as societies engage with the ethical, practical, and existential questions posed by genetic modification. The global dialogue is further complicated by the influence of international organizations and treaties, such as the Biological Weapons Convention, which aim to regulate the use of genetic technologies in warfare. These treaties reflect a collective effort to prevent the misuse of powerful technologies, but their effectiveness relies heavily on the willingness of individual nations to adhere to agreed norms and standards.

As the conversation continues, it is crucial to consider the diverse cultural landscapes that shape public opinion and policy. Understanding these reactions helps illuminate the broader societal implications of deploying CRISPR technology in warfare, highlighting the need for a nuanced and globally informed approach to this rapidly advancing field. This cultural mosaic sets the stage for examining the public perception and media influence on genetic modification in warfare, revealing the complex interplay between societal values and technological innovation.

As we delve deeper into the societal implications of CRISPR technology and its military applications, it is imperative to consider the broader impact on human rights and social dynamics. The introduction of genetically modified soldiers and the potential for genetic optimization in the general population raises profound questions about discrimination and integration in society.

The promise of genetic editing through CRISPR brings with it the possibility of eliminating hereditary diseases and enhancing human capabilities. However, this same promise also harbors the risk of exacerbating existing social inequalities. Historically, technological advancements have often led to new forms of social stratification. For instance, access to early computers and the internet created a digital divide, predominantly along socioeconomic lines. Similarly, access to genetic enhancements may become a privilege reserved for the wealthy, further entrenching social divides.

Discrimination based on genetic traits is not a new concept. In recent history, the eugenics movement, which sought to improve the genetic quality of human populations, led to forced sterilizations and other human rights abuses. This dark chapter serves as a cautionary tale for the modern era, highlighting the potential for misuse of genetic technologies. The question then arises: if we begin to enhance soldiers or civilians genetically, how do we ensure that it does not lead to a new form of genetic discrimination?

Another pressing concern is that the use of DNA databases by law enforcement agencies and the potential military use of gene-editing technologies like CRISPR share significant similarities in terms of ethical considerations, privacy concerns, and the challenges of regulating advanced biotechnologies. Both applications underscore the difficulty of containing powerful technologies once they become available and widely used.

Law enforcement agencies have increasingly utilized DNA databases from consumer genealogy websites such as Ancestry.com and 23andMe to solve cold cases and track down suspects. This practice gained widespread attention with the capture of the Golden State Killer in 2018, using DNA from a relative's genealogy profile. The Golden State Killer, also known as Joseph James DeAngelo, was responsible for a series of heinous crimes including murders, rapes, and burglaries that terrorized California from the 1970s to the 1980s. For decades, these crimes remained unsolved despite extensive investigations and the collection of various forms of evidence, including DNA samples from crime scenes.

The breakthrough came when investigators uploaded a DNA profile obtained from crime scene evidence to a public genealogy database called GEDmatch. Unlike commercial databases such as Ancestry.com and 23andMe, GEDmatch allows users to upload their raw genetic data obtained from different testing companies to find potential relatives. By comparing the DNA from the crime scene to the profiles available on GEDmatch, investigators were able to identify distant relatives of the suspect.

This familial search led to the construction of a family tree, narrowing down the pool of potential suspects based on age, location, and other factors. Eventually, Joseph James DeAngelo emerged as a likely candidate. To confirm his identity, investigators collected a discarded DNA sample from DeAngelo and matched it to the DNA evidence from the crime scenes, conclusively linking him to the crimes. DeAngelo was arrested in April 2018 and later pled guilty to multiple charges, bringing closure to numerous victims and their families.

The use of a genealogy database to solve this high-profile case demonstrated the power of DNA technology and its potential to solve cold cases. However, it also sparked significant debate about privacy and ethical considerations. Critics argued that individuals who uploaded their DNA to genealogy databases did not necessarily consent to their data being used in criminal investigations, raising concerns about informed consent and the potential for misuse. Proponents, on the other hand, highlighted the immense potential of such databases to bring justice in cases that had long been considered unsolvable.

This landmark case underscored the dual-edged nature of DNA technology: its ability to revolutionize law enforcement and criminal justice, while also posing serious questions about privacy, consent, and the ethical use of genetic information.

The potential military use of CRISPR for creating genetically enhanced soldiers poses a different set of challenges but shares the same overarching issue of controlling advanced technology. CRISPR can potentially be used to enhance physical and cognitive abilities, making soldiers stronger, faster, and more resilient. However, the ethical implications are profound. The line between therapeutic use and enhancement can blur, raising questions about consent, the potential for misuse, and long-term societal impacts. Once the technology is developed, it is challenging to prevent it from being used for purposes beyond its initial intent.

In both cases, issues of privacy and consent are paramount. With DNA databases, individuals might unknowingly have their genetic information used in ways they did not consent to. Similarly, soldiers might be subject to genetic enhancements without fully understanding or consenting to the long-term implications. Both technologies highlight the difficulty of regulation. DNA databases, initially intended for personal use, have become tools for law enforcement. Similarly, CRISPR, developed for medical and scientific advancements, could be appropriated for military use. Once these technologies are out in the open, containing their use becomes nearly impossible.

The ethical concerns are significant in both scenarios. Using DNA to solve crimes can prevent future offenses and bring closure to victims' families, but it also risks privacy violations and potential misuse. Military applications of CRISPR could enhance national security and soldier safety but might also lead to a new arms race and ethical dilemmas about human enhancement. Both technologies have the potential for unintended consequences. Misuse of genetic databases could lead to wrongful convictions or privacy breaches. Misuse of CRISPR in the military could lead to unforeseen health issues, ethical concerns, and societal divisions.

The comparison between law enforcement's use of DNA databases and the potential military use of CRISPR underscores the complexities and ethical dilemmas inherent in advanced biotechnologies. Both highlight the need for robust regulatory frameworks, ethical guidelines, and ongoing public dialogue to navigate the challenges posed by these powerful tools. The difficulty in keeping such technologies out of the hands of law enforcement and the military once they become available emphasizes the importance of proactive measures to ensure their responsible and ethical use.

International human rights organizations have already started to voice concerns about the ethical implications of genetic editing. The United Nations' Universal Declaration on the Human Genome and Human Rights explicitly states that genetic engineering should not lead to discrimination against individuals or groups. This landmark document, adopted in 1997, was a proactive step to ensure that advancements in genetic science respect human dignity and rights. Article 6 of the Declaration clearly articulates that no one shall be subjected to discrimination based on their genetic characteristics, emphasizing the need to prevent genetic information from being used to stigmatize or disadvantage individuals.

Despite these principles, enforcing them in a world where genetic enhancements become commonplace presents significant challenges. One major concern is the potential for genetic enhancements to create a new form of social inequality. If access to genetic enhancements is limited to those who can afford them, it could exacerbate existing social and economic disparities, leading to a society where the genetically enhanced have significant advantages over those who are not. This could affect various aspects of life, including education, employment, and healthcare, deepening the divide between different socioeconomic groups.

Moreover, there is the risk of genetic enhancements being used for eugenic purposes. History has shown the dangers of eugenics, where attempts to "improve" the human population through selective breeding led to gross human rights violations. With genetic engineering, there is a fear that similar ideologies could resurface, promoting the idea of creating "perfect" humans while marginalizing those who do not meet certain genetic criteria. This could lead to widespread discrimination and a loss of genetic diversity, which is crucial for the resilience and adaptability of the human species.

Another enforcement challenge is the global nature of genetic technology development. Countries may have varying regulations and ethical standards regarding genetic engineering, making it difficult to establish a unified approach to prevent discrimination. For instance, while some countries may strictly regulate genetic enhancements to ensure equity and ethical compliance, others might adopt more lenient policies to attract scientific research and investment. This disparity can lead to "genetic tourism," where individuals travel to countries with fewer regulations to obtain genetic enhancements, further complicating the enforcement of universal ethical standards.

Genetic tourism, like the trend of Westerners traveling to other countries for medical treatment where it is cheaper and less regulated, reflects the global disparities in healthcare regulation, cost, and accessibility. Both phenomena highlight how individuals seek out medical procedures in countries with less stringent regulations or lower costs, often driven by financial constraints or the desire to access treatments that may not be available or are heavily regulated in their home countries.

Genetic tourism involves individuals traveling to countries with more permissive regulations to undergo genetic modifications, such as CRISPR-based enhancements or treatments. These countries may offer advanced genetic interventions that are either not approved or are in the experimental stages in more regulated regions like the United States or Europe. The motivations for genetic tourism can include access to cutting-edge treatments that have not yet been approved by regulatory bodies in their home countries, lower costs in less regulated countries that make these procedures more accessible, and the desire to evade regulations for genetic interventions considered ethically or legally contentious in their home countries.

Similarly, medical tourism sees Westerners traveling to other countries to access medical treatments at a fraction of the cost they would pay domestically. Countries like India, Thailand, and Mexico have become popular destinations for procedures ranging from cosmetic surgery to major operations. The driving factors include lower costs due to lower labor and administrative expenses, quicker access to medical services without long waiting lists, and the availability of treatments that might not be accessible at home due to regulatory restrictions or limited availability.

Both genetic and medical tourism exploit the variations in regulatory environments. In countries with less stringent regulations, clinics and hospitals can offer treatments that are either not yet approved or are considered experimental elsewhere. This regulatory flexibility attracts patients willing to take the risks associated with less oversight. Cost is a significant driver for both types of tourism, as high healthcare costs in Western countries push patients to seek more affordable options abroad. This is particularly evident in medical tourism, where the cost of surgeries, dental work, and even routine procedures can be dramatically lower. Similarly, the high cost of advanced genetic treatments in regulated markets can lead to a search for more affordable options in countries with less stringent economic and regulatory constraints.

    Both forms of tourism raise ethical and safety concerns. In genetic tourism, the potential for unforeseen genetic consequences and the lack of long-term studies on safety and efficacy pose significant risks. In medical tourism, the quality of care can vary widely, and there is often less recourse for patients if something goes wrong. Both types of tourism can result in substandard outcomes due to the variability in medical standards and practices across countries.

    Medical tourism can place a strain on local healthcare systems in destination countries, where resources may be diverted to cater to foreign patients at the expense of local populations. Similarly, the influx of genetic tourists can lead to ethical dilemmas and resource allocation issues, particularly in countries that may lack robust regulatory frameworks to manage the complexities of advanced genetic treatments. Both phenomena underscore the global health disparities that drive individuals to seek care outside their home countries. They highlight the uneven distribution of medical and genetic technologies and the challenges in ensuring equitable access to these advancements worldwide.

Genetic tourism and medical tourism for cheaper, less regulated treatments share common themes of cost-saving, regulatory evasion, and access to advanced or timely care. Both reflect broader issues in global healthcare, including disparities in access, ethical considerations, and the challenges of maintaining consistent standards across different regulatory environments. As genetic technologies continue to advance, the parallels between these two forms of medical travel will likely become more pronounced, necessitating a nuanced approach to regulation, ethics, and patient safety on a global scale.

In a world where advanced genetic technologies like CRISPR are becoming increasingly accessible, the prospect of terrorist groups exploiting lax regulations in certain regions to create genetically modified super terrorists is a growing concern. These groups, driven by extremist ideologies and willing to use any means necessary to achieve their goals, could potentially travel to countries with minimal oversight on genetic editing to develop enhanced operatives. These modifications could include increased physical strength, enhanced cognitive abilities, accelerated healing, and resistance to extreme environmental conditions, making these super terrorists formidable adversaries.

The potential for such developments is particularly alarming given the relative ease with which CRISPR technology can be acquired and utilized. Unlike traditional weapons, genetic editing tools do not require large infrastructure or conspicuous materials, making them easier to conceal and transport. Terrorist groups could establish clandestine laboratories in countries with weak regulatory frameworks, away from the prying eyes of international oversight. These laboratories could become breeding grounds for developing genetically enhanced operatives who are more resilient, lethal, and capable of carrying out complex missions with unprecedented efficiency.

The implications of genetically modified super terrorists are profound. Enhanced physical abilities would allow these individuals to endure grueling conditions, carry out prolonged operations, and engage in combat with superior endurance and strength. Cognitive enhancements could lead to operatives with heightened strategic thinking, better decision-making under pressure, and an ability to manipulate information and technology more effectively. Moreover, accelerated healing and resistance to environmental stresses would make these operatives harder to neutralize and more adaptable to various operational scenarios.

The international community faces significant challenges in addressing this potential threat. The disparity in regulatory standards across countries creates loopholes that can be exploited by malicious actors. While some nations have stringent controls on genetic research and modifications, others lack the resources or political will to enforce comprehensive regulations. This patchwork of regulatory environments provides fertile ground for terrorist groups seeking to harness the power of genetic editing for nefarious purposes.

Imagine this fictitious scenario that could become a reality within the near future.

In the dimly lit basement of an abandoned warehouse on the outskirts of a bustling city in Eastern Europe, a group of four men huddled around a flickering laptop screen. The men, known only by their code names—Jamal, Karim, Fadil, and Tariq—were members of a highly secretive terrorist organization called Al-Qadim. Their mission: to undergo genetic modifications and become the world's first super terrorists, capable of executing a devastating attack on the United States.

The leader of Al-Qadim, an enigmatic figure known only as The Architect, had meticulously planned every detail of Operation Genesis. He had identified a small, war-torn country with virtually no regulation on genetic research as the ideal location for the modifications. The country's lack of oversight and desperate need for funding made it easy for The Architect to bribe local officials and secure the services of Dr. Viktor Koval, a rogue scientist with a dark reputation for his work in genetic engineering.

Jamal, Karim, Fadil, and Tariq were chosen for their unique skills and unwavering loyalty to the cause. Jamal, a former elite commando, was to receive enhancements to his strength and endurance. Karim, an expert hacker, would undergo modifications to his cognitive abilities, allowing him to process information at an unprecedented speed. Fadil, a master of disguise and infiltration, was to gain accelerated healing and resistance to toxins. Tariq, a sharpshooter with unmatched precision, would receive enhancements to his eyesight and reflexes.

The journey to the clandestine laboratory was fraught with countless signs of danger. The four operatives traveled under false identities, moving through a network of safe houses and using encrypted communication to avoid detection. After a few weeks of covert travel, they finally arrived at the decrepit facility hidden deep in the mountains, where Dr. Koval awaited them with his team of scientists.

Dr. Koval's laboratory was a maze of advanced genetic equipment and makeshift living quarters. The walls were lined with cages containing genetically modified animals—eerie reminders of the experimental nature of their mission. The operatives were each subjected to a series of grueling tests and procedures. Over the course of several months, their bodies were transformed at the genetic level. Pain and uncertainty were constant companions, but the promise of unparalleled power kept them focused on their objective.

As the enhancements took hold, the operatives began to notice the changes. Jamal could lift weights that would crush an ordinary man. Karim's mind became a supercomputer, capable of hacking the most secure systems in seconds. Fadil's wounds healed almost instantly, and he could withstand doses of poison that would kill a normal human. Tariq's vision sharpened to the point where he could see in the dark and hit targets with pinpoint accuracy from incredible distances.

With their transformations complete, the four super terrorists were ready to execute the final phase of Operation Genesis. They returned to their base of operations, where The Architect briefed them on their mission. The plan was to attack a high-profile target in the United States, demonstrating their newfound abilities and striking fear into the hearts of their enemies.

The target was the Freedom Tower in New York City. The mission was meticulously planned: Jamal would lead the physical assault, neutralizing security and creating chaos. Karim would hack into the building's security systems, disabling cameras and alarms. Fadil would infiltrate the tower, planting explosives at strategic points. Tariq would provide overwatch, taking out key personnel from a distance to ensure the operation's success.

On a cold, clear morning, the four operatives slipped into the United States, blending in with the city's bustling crowds. They moved with precision and coordination, each carrying out their assigned tasks with deadly efficiency. The Freedom Tower, a symbol of resilience and hope, was about to become the stage for a new kind of warfare.

Jamal's enhanced strength allowed him to overpower security guards effortlessly, while Karim's hacking skills plunged the building into darkness. Fadil's healing abilities enabled him to survive an extended firefight that led to many casualties. He was able to continue his mission, planting explosives that would bring the tower down. Tariq's enhanced vision and reflexes ensured that no one could interfere with their plan.

As the final moments of the operation unfolded, the team executed their escape plan, disappearing into the chaos they had created. The Freedom Tower stood no chance against the combined might of the genetically enhanced operatives. The explosion was devastating, leaving a city and a nation in shock.

Operation Genesis was a grim demonstration of the potential for genetic modifications to create a new breed of super terrorists. The attack on the Freedom Tower served as a stark warning of the dangers of unregulated genetic engineering, showing how easily such technology could be turned against society. The world was forced to confront the terrifying reality that the future of warfare had arrived, bringing with it unprecedented ethical and security challenges.

Addressing this threat requires a coordinated global response. Strengthening international regulations on genetic editing, increasing cooperation between intelligence agencies, and investing in advanced surveillance technologies are crucial steps. Moreover, raising awareness about the risks associated with unregulated genetic modifications and promoting ethical standards in scientific research can help mitigate the potential misuse of this powerful technology. As genetic editing continues to evolve, the international community must remain vigilant to ensure that these advancements are not co-opted by those who seek to use them for terror and destruction.

Additionally, the rapid pace of technological advancement in genetics often outstrips the development of corresponding legal and ethical frameworks. Legislators and policymakers frequently find themselves playing catch-up, trying to address the implications of new technologies after they have already been introduced. This lag can result in a period where genetic enhancements are available without adequate oversight, increasing the risk of misuse and discrimination.

Public perception and acceptance of genetic enhancements also play a crucial role. If society views genetic enhancements as a norm or even as a necessity to compete in various fields, it could pressure individuals to undergo enhancements against their will or better judgment. This societal pressure can lead to a new form of coercion, where the choice not to enhance oneself or one's children results in significant disadvantages.

While the United Nations' Universal Declaration on the Human Genome and Human Rights sets a vital ethical framework to prevent discrimination based on genetic characteristics, the practical enforcement of these principles faces numerous challenges. Addressing these challenges requires a concerted global effort to develop robust legal, ethical, and regulatory measures that can keep pace with technological advancements, ensuring that genetic engineering serves the common good without compromising human rights and dignity.

Moreover, the integration of genetically modified individuals into society poses unique challenges. On one hand, enhanced individuals might be perceived as superior, potentially leading to a new elite class. On the other hand, those who are not genetically modified could face stigmatization and marginalization. This dichotomy could disrupt social cohesion and lead to conflict.

The psychological impact on individuals themselves cannot be overlooked. For soldiers enhanced through CRISPR, the transition back to civilian life may be fraught with difficulties. Their enhanced abilities, while advantageous on the battlefield, could become sources of alienation in everyday interactions. This phenomenon has parallels with the struggles faced by veterans of conventional wars, who often experience difficulties reintegrating into civilian life due to their unique experiences and skills.

Ethicists argue that to mitigate these risks, there must be robust policies and regulations in place to ensure fair and equitable access to genetic technologies. Additionally, there needs to be a concerted effort to foster a societal ethos that values genetic diversity and recognizes the inherent dignity of all individuals, regardless of their genetic makeup. This requires a multidisciplinary approach, involving policymakers, scientists, ethicists, and the public, to navigate the complex terrain of genetic advancements.

Chapter 10.

Future Prospects and Speculations

As we delve deeper into the future possibilities of CRISPR technology, we must consider the speculative advancements that could reshape not only warfare but also society at large. The potential for CRISPR to revolutionize genetic engineering seems boundless, pushing the boundaries of what we currently deem possible.

One of the most intriguing aspects of CRISPR is its ability to target and modify specific genes with remarkable precision. This has already opened doors to potential cures for genetic disorders and advancements in agricultural biotechnology. Looking ahead, scientists are exploring how CRISPR could be used to enhance human capabilities beyond natural limitations. For instance, researchers are investigating the possibility of using CRISPR to enhance physical traits such as muscle strength, endurance, and even cognitive abilities. These enhancements could create individuals with superior physical and mental faculties, tailored for specific tasks or environments.

In the realm of medicine, CRISPR's future applications are particularly exciting. Researchers are developing CRISPR-based therapies to target and eradicate cancers more effectively. By editing the genes of immune cells to better recognize and attack cancer cells, we could see a significant shift in cancer treatment paradigms. Additionally, the potential to edit the human germline—the genetic material passed from one generation to the next—raises the prospect of eliminating hereditary diseases altogether. While this prospect is ethically contentious, the scientific community continues to debate its feasibility and moral implications.

Moreover, CRISPR could play a pivotal role in addressing global health challenges. For instance, the technology is being explored as a tool to combat vector-borne diseases such as malaria and dengue fever. By genetically modifying mosquitoes to reduce their ability to transmit these diseases, CRISPR could significantly lower infection rates and save millions of lives annually.

As we speculate on these advancements, it's essential to consider the broader implications. The possibility of using CRISPR for human enhancement raises significant ethical questions. What does it mean to be human if we can artificially enhance our physical and mental capabilities? The potential for creating a divide between genetically enhanced individuals and those who remain unmodified could exacerbate social inequalities, leading to new forms of discrimination and societal tension.

Furthermore, the concept of genetic enhancement is not limited to the battlefield or medical field. In agriculture, CRISPR could revolutionize food production by creating crops that are more resistant to pests, diseases, and climate change. This could address global food security issues and reduce the environmental impact of farming practices. However, the widespread adoption of genetically modified organisms (GMOs) remains a contentious topic, with ongoing debates about their safety and ethical implications.

The pace of technological advancement in genetic engineering is accelerating, and with it comes a host of new possibilities and challenges. As we look towards a future where CRISPR and other genetic technologies become more integrated into various aspects of life, it is crucial to engage in thoughtful and informed discussions about their potential impacts. Balancing the promise of these advancements with the ethical and societal considerations they entail will be key to ensuring that we navigate this new frontier responsibly.

Transitioning from speculative advancements in genetic engineering, it becomes evident that the implications extend far beyond the laboratory. The integration of CRISPR technology into military applications presents a particularly complex set of challenges and opportunities that warrant careful consideration.

This intersection of cutting-edge science and global security highlights the need for robust regulatory frameworks and ethical guidelines to govern the use of genetic technologies in warfare. As we proceed to explore the geopolitical ramifications of CRISPR-enhanced soldiers, the importance of international cooperation and regulation cannot be overstated.

In examining the future prospects of CRISPR technology, particularly its application in military contexts, it is imperative to address the critical need for robust policy recommendations. The rapid advancement of genetic editing tools, such as CRISPR, brings with it profound ethical, legal, and social implications. Therefore, proactive regulation is essential to prevent potential misuse and to ensure that these powerful technologies are deployed responsibly.

The potential benefits of CRISPR in medicine and agriculture are immense, but when it comes to military applications, the stakes are even higher. The creation of genetically modified soldiers, or "super-soldiers," presents a unique set of challenges that require careful consideration. One of the primary concerns is the ethical dilemma of enhancing human beings for combat purposes. Historical precedents, such as the eugenics movement, underscore the dangers of pursuing genetic "perfection" and highlight the need for stringent ethical guidelines.

To regulate the military use of CRISPR, international cooperation is crucial. Similar to the frameworks established for nuclear non-proliferation, a global agreement on genetic editing in warfare should be pursued. Such an agreement would ideally involve key international bodies like the United Nations and the World Health Organization. These organizations could help draft and enforce regulations that prohibit unethical genetic modifications and ensure transparency in military research.

National governments must also play a pivotal role. Countries should develop comprehensive legislation that outlines the permissible scope of genetic research within their military programs. This legislation should mandate strict oversight and include provisions for regular audits and reviews by independent bodies. For example, the United States could expand the jurisdiction of the Department of Defense's Inspector General to include oversight of genetic research projects. Similarly, other nations could establish equivalent bodies to monitor compliance with international standards.

Transparency is another critical aspect of regulation. Military research in genetic editing must be subject to public scrutiny to some extent. While certain details may need to remain classified for national security reasons, a balance must be struck to ensure that the public remains informed about the ethical considerations and potential risks involved. Implementing transparency measures, such as periodic public reports and open forums for discussion, can help build public trust and ensure accountability.

Additionally, the integration of bioethical education in military training programs is essential. Military personnel, particularly those involved in genetic research and its applications, should receive comprehensive training on the ethical implications of their work. This training would help cultivate a culture of ethical responsibility and ensure that decisions regarding genetic modifications are made with a thorough understanding of their potential consequences.

Finally, the rapid pace of technological advancement necessitates a flexible regulatory framework that can adapt to new developments. Policymakers must remain vigilant and ready to update regulations as new ethical, legal, and social challenges emerge. This adaptability will be crucial in maintaining the delicate balance between fostering innovation and preventing misuse.

The conversation surrounding the regulation of CRISPR technology in military applications is complex and multifaceted. It demands a collaborative approach that involves international bodies, national governments, and the public. By establishing comprehensive regulations and promoting ethical awareness, we can harness the potential of CRISPR while safeguarding against its potential dangers. As we continue to explore the future of genetic engineering, these policy recommendations provide a foundation for responsible and ethical advancement.

As we look beyond regulatory frameworks, it is equally important to consider the broader societal implications of genetic modifications in the military. The integration of CRISPR technology not only affects the soldiers who are directly modified but also has far-reaching consequences for society as a whole. The cultural and societal impacts of these advancements will be the focus of our next section, where we will delve into how different cultures might react to genetically modified soldiers and the broader societal changes that could ensue.

Imagine a battlefield where soldiers are no longer constrained by the limits of human biology. Enhanced strength, faster reflexes, and heightened cognitive abilities could turn the tide of any conflict. However, the use of CRISPR to achieve such enhancements brings forth ethical dilemmas that must be carefully weighed. The ability to manipulate the human genome, while promising in its potential to eradicate diseases, also opens the door to unintended consequences and moral quandaries.

One of the primary concerns is the concept of consent. Can a soldier truly consent to genetic modifications that will fundamentally alter their physiology and possibly their identity? This issue is compounded when considering the hierarchical nature of the military, where orders are followed without question. The possibility of coercion or undue pressure to undergo genetic enhancements cannot be dismissed lightly. History provides cautionary tales; the eugenics movement of the early 20th century, which aimed to create a "better" human race, ultimately led to gross human rights violations.

Furthermore, the long-term effects of genetic modifications are still largely unknown. While CRISPR technology has shown promise in laboratory settings, the real-world applications, especially under the stresses of combat, remain untested. The potential for unforeseen genetic disorders or other health complications poses significant risks not only to the individuals involved but also to military readiness and morale.

The integration of artificial intelligence into genetic optimization adds another layer of complexity. As AI systems increasingly dictate the parameters of genetic enhancements, the autonomy of individual soldiers could be significantly undermined. AI's role in optimizing human traits may lead to a homogenized force where diversity, a critical element in human resilience and adaptability, is sacrificed for perceived efficiency. This shift towards a more uniform human profile in military ranks could have far-reaching implications, including the erosion of individual freedoms and the suppression of unique human qualities that are often the source of innovation and creativity.

Imagine this possibility. The integration of CRISPR gene editing with brain-computer interfaces (BCIs) like Neuralink presents a formidable leap in the enhancement of military soldiers, combining physical and cognitive advancements in unprecedented ways. CRISPR technology, with its ability to precisely edit genes, offers the potential to enhance physical attributes such as strength, endurance, and resistance to diseases. When coupled with BCIs, which enable direct communication between the brain and external devices, the possibilities for creating super soldiers extend far beyond mere physical enhancements.

On the positive side, this combination could lead to soldiers with superior physical abilities and enhanced cognitive functions. CRISPR could be used to improve muscle efficiency and bone density, allowing soldiers to carry heavier loads and endure longer missions without fatigue. Simultaneously, BCIs like Neuralink could provide real-time data integration, allowing soldiers to process vast amounts of information quickly and make split-second decisions with unparalleled accuracy. This would enable soldiers to operate complex machinery, pilot drones, and manage battlefield logistics with enhanced precision and efficiency. Moreover, the ability to directly interface with communication systems could lead to improved coordination and execution of military operations, potentially reducing the risk of human error and increasing overall mission success.

However, the integration of CRISPR and BCIs also carries significant negative implications, particularly in the realms of brainwashing and mind control. The same technologies that enhance cognitive abilities can also be exploited to exert control over a soldier's mind. With direct access to neural pathways, there is a potential risk that BCIs could be used to manipulate thoughts, emotions, and behaviors. This raises profound ethical concerns about autonomy and free will, as soldiers could be subjected to programming or conditioning that overrides their personal agency.

Furthermore, the combination of genetic enhancements and neural interfaces could lead to the development of soldiers who are more susceptible to psychological manipulation. The ability to edit genes that influence brain function, combined with direct neural input, could theoretically be used to enforce obedience and suppress dissent, effectively creating a class of super soldiers who are not only physically superior but also mentally conditioned to follow orders without question. This level of control could be exploited by unethical leaders, leading to potential human rights abuses and the erosion of individual freedoms.

The implications for privacy and security are also significant. With BCIs, there is a risk of external hacking or cyber attacks, where adversaries could gain control over soldiers' neural interfaces, turning them into unwitting agents of sabotage. The potential for misuse in espionage and warfare is immense, as enemy forces could deploy these technologies to disrupt military operations and create internal chaos.

While the combination of CRISPR gene editing and brain-computer interfaces offers exciting possibilities for enhancing military capabilities, it also presents severe ethical, psychological, and security challenges. The potential for creating a new breed of super soldiers comes with the responsibility to address these risks and ensure that advancements in biotechnology and neural engineering are used to protect and enhance human dignity, rather than undermine it. The future of warfare will require careful consideration of these technologies' implications, balancing the quest for superiority with the imperative to uphold ethical standards and human rights.

    The future of warfare, shaped by the dual forces of genetic modification and artificial intelligence, holds both promise and peril. As we stand on the brink of this new era, it is incumbent upon us to navigate these uncharted waters with caution, guided by a commitment to ethical principles and a deep respect for human dignity. Only by doing so can we hope to harness the benefits of these technologies while safeguarding against their potential to harm.

Conclusion:

Preparing for a New Frontier

As we have journeyed through the complexities and possibilities of CRISPR technology and its potential applications in modern warfare, it's important to reflect on the key points that have shaped our understanding. The advent of CRISPR has revolutionized genetic editing, allowing for unprecedented precision in altering DNA. This innovation, initially hailed for its potential to eradicate genetic diseases and improve human health, has now found a place in military ambitions, raising both hopes and significant ethical concerns.

The history and development of CRISPR have been marked by rapid advancements and groundbreaking discoveries. Originating from the study of bacterial immune systems, CRISPR-Cas9 has transformed from a biological curiosity to a powerful tool with vast implications. Scientists like Jennifer Doudna and Emmanuelle Charpentier, who were instrumental in harnessing this technology, envisioned its use primarily in medicine and agriculture. However, its potential to enhance human capabilities quickly attracted the interest of military organizations worldwide.

The military's fascination with genetic engineering is not entirely new. Historical attempts to enhance soldier performance, from amphetamines in World War II to more recent research into cognitive and physical augmentation, set the stage for the integration of CRISPR technology. The allure of creating "super-soldiers" with enhanced strength, endurance, and cognitive abilities is compelling, yet it brings forth a myriad of ethical dilemmas. The possibility of engineering humans for warfare challenges our fundamental understanding of what it means to be human and raises profound moral questions about the extent to which we should alter our biology.

Throughout this exploration, we have also examined the technical and ethical challenges that accompany the use of CRISPR in military contexts. The precision of CRISPR technology, while remarkable, is not without its risks. Off-target effects, where unintended parts of the genome are altered, present significant safety concerns. Furthermore, the ethical implications of creating genetically modified soldiers cannot be understated. The potential for misuse, coercion, and the violation of individual rights are serious considerations that must be addressed.

As we look at the geopolitical landscape, it becomes evident that the use of CRISPR in warfare could trigger a new arms race. Countries vying for superiority in genetic enhancements may invest heavily in this technology, leading to an escalation in military tensions. The absence of comprehensive international regulations governing genetic modifications in warfare exacerbates these concerns, highlighting the need for robust policies and oversight.

The integration of AI in optimizing genetic traits introduces another layer of complexity. While AI can enhance the efficiency and precision of genetic modifications, it also poses risks to human autonomy. The potential for AI to dictate reproductive choices and prioritize certain traits over others threatens to undermine genetic diversity and individual freedoms. This intersection of AI and genetic engineering calls for a careful examination of the ethical boundaries and societal impacts of these technologies.

In considering the broader societal impacts, the division between genetically optimized individuals and those deemed "imperfect" mirrors existing societal inequalities. This division could lead to increased discrimination, social unrest, and resistance against oppressive systems. The historical precedent of eugenics serves as a stark warning of the dangers of striving for a "perfect" population, reminding us of the atrocities that can arise from such pursuits.

As we draw these reflections to a close, it is clear that the promise of CRISPR technology is both vast and fraught with challenges. The potential benefits in health and medicine are immense, but the application of this technology in warfare demands careful scrutiny and ethical consideration. The future of

CRISPR in military contexts will depend on our ability to navigate these complexities and develop frameworks that ensure its responsible use. This journey through the possibilities and perils of genetic engineering in warfare underscores the need for vigilance, ethical integrity, and a commitment to preserving our shared humanity.

As we look ahead to the future of CRISPR in warfare, we find ourselves standing at a crossroads where technology and ethics intersect. The advancements we have discussed throughout this book paint a picture of incredible possibilities and profound risks. CRISPR's ability to precisely edit genes opens up unprecedented opportunities to enhance human capabilities, but it also poses significant ethical and moral dilemmas.

One of the most pressing concerns is the potential for misuse. While CRISPR technology can be harnessed for the betterment of humanity, its application in military contexts raises the specter of an arms race centered not on nuclear weapons, but on genetically enhanced soldiers. History provides ample warnings about the dangers of such pursuits. For example, the eugenics movement of the early 20th century, though based on flawed science, led to widespread human rights abuses. This historical precedent underscores the need for stringent ethical guidelines and robust international regulations to govern the use of genetic editing in the military.

Moreover, the potential for unintended consequences cannot be overstated. Genetic modifications that seem beneficial in the short term might have unforeseen long-term effects. The complexity of human genetics means that changes intended to enhance certain traits could inadvertently introduce new vulnerabilities or health issues. For instance, attempts to increase physical strength or cognitive abilities could result in unforeseen physiological or psychological side effects, complicating the ethical landscape further.

Additionally, the societal implications of deploying CRISPR-enhanced soldiers are profound. Such advancements could exacerbate existing inequalities and create new forms of discrimination. In a world where genetic enhancements become a measure of military might, nations with advanced CRISPR capabilities could dominate those without, leading to geopolitical instability. This scenario calls for a careful consideration of how to maintain a balance of power and ensure that such technologies do not deepen global divides.

The promise of CRISPR is not without its bright spots. Its potential to eradicate genetic diseases and improve human health is a testament to its power. Yet, as we consider its application in warfare, we must proceed with caution. The ethical questions it raises demand thorough and thoughtful deliberation. As a society, we must grapple with what it means to be human and how far we are willing to go in our quest for improvement.

Ultimately, the future of CRISPR in warfare will depend on the decisions we make today. It will require a collective effort from scientists, policymakers, ethicists, and the public to navigate this complex terrain. The path we choose will shape not only the nature of future conflicts but also the essence of humanity itself. As we move forward, we must strive to harness the potential of CRISPR responsibly, ensuring that its use in warfare does not compromise our ethical standards or humanity's core values.

Our journey through the possibilities and perils of CRISPR in warfare has brought us to a pivotal moment. The choices we make now will reverberate through generations to come, influencing not just the future of military technology but the very fabric of human society. As we close this exploration, we must carry forward the lessons learned and continue the dialogue on how to ethically and effectively integrate such powerful technology into our world.

As we have explored the numerous dimensions of CRISPR technology and its potential to revolutionize warfare, the implications extend far beyond the battlefield. This technology touches the very essence of what it means to be human and the structure of our societies. In light of these profound implications, it is imperative that we, as a global community, take decisive and thoughtful actions to navigate the ethical, scientific, and social challenges presented by genetic engineering in military contexts.

Policymakers stand at the forefront of this crucial endeavor. They must craft regulations that ensure the responsible use of

CRISPR technology while preventing its misuse. The current international laws, such as the Biological Weapons Convention, do not explicitly address the use of genetic modification in warfare, creating a regulatory gap that needs urgent attention. By establishing clear guidelines and frameworks, policymakers can help prevent an unchecked arms race fueled by genetic enhancements and ensure that military applications of CRISPR adhere to ethical standards. The potential for misuse is significant, and without stringent oversight, the consequences could be dire.

Scientists, too, have a pivotal role to play. The research community must prioritize transparency and ethical considerations in their work. This includes rigorous peer review and public dissemination of findings to foster an informed dialogue about the capabilities and limitations of CRISPR technology. Collaborative efforts between scientists, ethicists, and policymakers can help develop robust safeguards that address both the potential benefits and the ethical dilemmas associated with genetic engineering. Historical precedents, such as the misuse of eugenics, highlight the importance of maintaining a vigilant and ethical approach to scientific advancements.

The public must also be engaged in this conversation. Public opinion and societal values will significantly influence the direction and acceptance of genetic technologies. Educational initiatives are essential to equip people with the knowledge to understand and critically evaluate the implications of CRISPR. Media representation plays a crucial role in shaping public perception; therefore, it is vital that the information disseminated is accurate, balanced, and free from sensationalism. By fostering a well-informed public, we can cultivate a society that is capable of making thoughtful decisions about the future of genetic engineering.

Moreover, the preservation of genetic diversity is a cornerstone of our humanity. As we stand on the brink of potentially altering the human genome, it is crucial to remember the lessons of history and the intrinsic value of diversity. The quest for a "perfect" population can lead to the erosion of human rights and individual freedoms, as seen in past atrocities committed in the name of eugenics. We must strive to protect the genetic diversity that makes us resilient and unique.

Ultimately, the future of CRISPR technology in warfare is not predetermined. By taking proactive and collaborative steps, we can steer its development in a direction that benefits humanity as a whole. The convergence of policy, science, and public engagement will be essential in navigating this complex landscape. Together, we can ensure that the use of CRISPR in military contexts is guided by ethical principles and a commitment to preserving human dignity.

In reflecting on these points, it becomes clear that our collective actions today will shape the future of genetic engineering. It is our responsibility to ensure that this powerful technology is used to enhance, rather than diminish, the human experience.

Sources

These sources have been integral in exploring the multifaceted implications of CRISPR technology and its potential application in creating genetically enhanced soldiers. They provide a comprehensive view of the scientific advancements, ethical considerations, and security challenges associated with this powerful technology.

1. Ratcliffe, J. (2020). "China's Super Soldiers: A Threat to Global Security". Wall Street Journal. Retrieved from WSJ
2. South China Morning Post. (2023). "Chinese team behind extreme animal gene experiment says it may lead to super soldiers who survive". Retrieved from SCMP
3. NBC News. (2020). "China has done human testing to create biologically enhanced super soldiers". Retrieved from NBC News
4. Kania, E., & VornDick, W. (2019). "Weaponizing Biotech: How China's Military Is Preparing for a 'New Domain of Warfare'". Retrieved from Defense One
5. Singer, P. W., & Cole, A. (2019). "Wired for War: The Robotics Revolution and Conflict in the 21st Century". Penguin Press.

6. Garreau, J. (2005). "Radical Evolution: The Promise and Peril of Enhancing Our Minds, Our Bodies—and What It Means to Be Human". Doubleday.
7. Center for a New American Security. (2017). "CRISPR and National Security: Threats and Opportunities". Retrieved from CNAS
8. Scientific American. (2017). "The Future of CRISPR: Editing the Invisible". Retrieved from Scientific American
9. U.S. Department of Defense. (2020). "DARPA's Biotechnological Initiatives". Retrieved from DARPA
10. The Forteana Forums. (2023). Discussions on genetic enhancements and military applications. Retrieved from Forteana Forums
11. Wynn, C. (2018). "The Implications of Genetic Engineering in Modern Warfare". Journal of Military Ethics.
12. Maron, D. F. (2017). "DARPA's Biotech Chief Says 2017 Will Blow Our Minds". Scientific American. Retrieved from Scientific American
13. U.S. Department of Veterans Affairs. (2015). "PTSD in Iraq and Afghanistan Veterans". Retrieved from VA Public Health
14. Invisible Wounds of War: Psychological and Cognitive Injuries, Their Consequences, and Services to Assist Recovery. (2008). RAND Corporation.
15. Reisman, M. (2016). "PTSD Treatment for Veterans: What's Working, What's New, and What's Next". Pharmacy and Therapeutics, 41(10), 623.

16. Dursa, E. K., et al. (2014). "Prevalence of a Positive Screen for PTSD Among OEF/OIF and OEF/OIF-Era Veterans in a Large Population-Based Cohort". Journal of Traumatic Stress, 27(5), 542.
17. Boeke, J. D., Church, G., Hessel, A., Kelley, N. J. (2016). "Genome Project-write: A Grand Challenge Using Synthesis, Gene Editing and Other Technologies to Understand, Engineer and Test Living Systems". Retrieved from Engineering Biology Center
18. Yong, E. (2017). "Now That We Can Read the Genomes, Can We Write Them?". The Atlantic. Retrieved from The Atlantic
19. Bielitzki, J., & Garreau, J. (2005). "Perfecting the Human". Retrieved from Lycaeum
20. Tanielian, T., & Jaycox, L. H. (2008). "Invisible Wounds of War: Psychological and Cognitive Injuries, Their Consequences, and Services to Assist Recovery". RAND Corporation.
21. BBC News. (2021). "China gene-edited babies: Who is He Jiankui?". Retrieved from BBC
22. Duke University. "Ethics and Policy: CRISPR and Gene Editing". Retrieved from Duke University
23. Futurism. "Scientist Tweaks Genetically Altered Super Soldiers". Retrieved from Futurism
24. RAND Corporation. (2021). "Gene Editing and National Security: Risks and Opportunities". Retrieved from RAND

25. The Washington Quarterly. (2020). "Gene Editing and Security: Risks, Oversight, and Governance". Retrieved from Tandfonline
26. ResearchGate. (2021). "From CRISPR Babies to Super Soldiers: Challenges and Security Threats Posed by CRISPR". Retrieved from ResearchGate
27. North Carolina Central University. "Legal and Ethical Implications of CRISPR Technology". Retrieved from NCCU
28. Synergia Foundation. "Transhumanism and Genetic Super Soldiers". Retrieved from Synergia Foundation

www.ingramcontent.com/pod-product-compliance
Lightning Source LLC
Chambersburg PA
CBHW071826210526
45479CB00001B/11